I0511029

# *The Pelton Water Wheel*

*The Pelton System of Power*

*by Pelton Water Wheel Company*

***with an introduction by Roger Chambers***

## *Introduction*

I am pleased to present yet another title on Homesteading and Farm Life.

This volume is entitled "The Pelton Water Wheel" and was published in 1898.

The work is in the Public Domain and is re-printed here in accordance with Federal Laws.

As with all reprinted books of this age that are intended to perfectly reproduce the original edition, considerable pains and effort had to be undertaken to correct fading and sometimes outright damage to existing proofs of this title. At times, this task is quite monumental, requiring an almost total "rebuilding" of some pages from digital proofs of multiple copies. Despite this, imperfections still sometimes exist in the final proof and may detract from the visual appearance of the text.

I hope you enjoy reading this book as much as I enjoyed making it available to readers again.

Roger Chambers

## IMPORTANT NOTICE.

The high reputation and extraordinary demand for PELTON WATER WHEELS in the home as well as foreign markets has led to MANY IMITATIONS and SOME INFRINGEMENTS of the various patents of this company.

Prospective purchasers are HEREBY WARNED that ALL SUCH INFRINGEMENTS will be VIGOROUSLY PROSECUTED, and that the USERS OF WHEELS so infringing, as well as THE MANUFACTURERS, will be held EQUALLY RESPONSIBLE. They are also reminded that in addition to incurring the liabilities above mentioned in dealing with UNAUTHORIZED MANUFACTURERS, wheels so obtained will in most cases prove to be ALTOGETHER INEFFICIENT AND INADEQUATE to the service required. Many cases have been brought to notice, more especially in foreign countries, where such attempts at imitation have proved TOTAL FAILURES, involving purchasers in SERIOUS LOSS AND TROUBLE.

Special attention is called to the fact that while PELTON WHEELS are sold through various agencies in all civilized countries, NO WHEELS OTHER THAN THOSE MADE BY THIS COMPANY, at either its SAN FRANCISCO OR NEW YORK WORKS, and bearing the trademark "PELTON WATER WHEEL," are genuine.

## ATLANTIC DEPARTMENT.

OF THE

# PELTON WATER WHEEL COMPANY,

143 LIBERTY STREET, NEW YORK, U. S. A.

THE PELTON WATER WHEEL COMPANY calls special attention to the fact that the ATLANTIC DEPARTMENT of this company—located at above address—is thoroughly equipped for furnishing full and reliable information in regard to any proposition for the utilization of water-powers by THE MOST MODERN, ECONOMIC, AND APPROVED METHODS; also as to its UNSURPASSED FACILITIES for the manufacture and prompt delivery at this point of water wheels and all appliances connected with works of this character, herein referred to.

Attention is also called to the fact that they furnish--of the best material and construction and at the lowest ruling prices—WATER PIPE—RIVETED SHEET STEEL AND LAP WELD—GATE VALVES AND FITTINGS—SHAFTING—GEARING—PULLEYS—SHEAVES—JOURNAL BEARINGS, as well as ALL MACHINERY AND ACCESSORIES connected with water-power installations.

The advantages this location affords for FAVORABLE FREIGHT RATES, as well as RAPID AND FREQUENT COMMUNICATION with all the markets of the world, are too apparent to enlarge upon. Such information as may be desired in reference to any proposition, from localities where this office is most accessible, will be furnished either by personal interview or inquiry by letter to address above given.

Parties writing for information will please confine their application to ONE OF THE OFFICES of this company, whichever may be most convenient in point of distance or facility of communication.

THE PELTON WATER WHEEL COMPANY.

PELTON WATER-WHEEL—IRON-MOUNTED TYPE.

# THE PELTON SYSTEM OF POWER.

## HIGHEST AWARD AT THE WORLD'S COLUMBIAN EXPOSITION.

"Countless wealth is being squandered in all the torrents and water-courses of the world."—*Prof. Ayrton, F. R. S.*

The use of water for purposes of power dates back to the early centuries, and, even with the crude and primitive means then available, was made to subserve many useful purposes. It is, however, only within a comparatively short time that it has come to be recognized as the most practicable and potent of all elemental forces, destined in the near future to do a large part of the world's work.

The advantages such a source of power offers are so many and obvious that even a reference to them seems superfluous. That these advantages have often been largely discounted by the use of appliances for utilization, either altogether unreliable or inadequate to the work intended, is unfortunately quite too common an experience.

The practise which has so long prevailed, of appropriating only the larger streams, with low heads, allowing the higher heads to go to waste, is attended with so many difficulties and such expense as to make a power so obtained often of questionable expediency.

The SYSTEM here presented offers so many advantages for the general utilization of all these sources of energy, everywhere abounding, that streams favorably situated for power purposes are now being eagerly sought for and appropriated.

By its use the entire force or highest head obtainable can be made available for all industrial purposes with a greatly reduced cost, wider range of application, and fuller adaptation to varying requirements, than has ever before been realized. Nothing in a mechanical way has so signally and quickly demonstrated its own usefulness as well as its right to the first place in hydraulic power appliances.

The transmission of power by electricity has opened up a new field and a wider range for the application of such sources of power, than has been possible by any previous methods, and the field is continually enlarging as the science advances. This most subtle and mysterious element has been brought under control and made subservient to industrial and commercial interests to an extent that can hardly be realized by those not fully conversant with the rapid and important developments of the past few years.

It is, however, in connection with water as an initial force that the highest development in this direction has been attained. Every stream or waterfall is a mine of energy that by means of this most simple appliance can be converted directly into useful effect with almost an entire absence of machinery, and made available for any desired purpose with a high degree of efficiency and comparatively small outlay.

Water that for thousands of years has run to waste, can now be transformed into electrical energy and carried long distances and made as useful as at the point where generated, and at small cost in most cases as compared with steam. The immeasurably vast resources of power available by this means open up in all directions new fields for enterprise, affording profitable employment for both labor and capital.

This subject is so intimately connected with water-power development that a very considerable space has been given it in this edition, which, with illustrations, descriptions, tables, and other related matter, it is hoped will be of interest to those contemplating installations of this character.

## GENERAL INFORMATION.

The power of the Pelton Wheel does not depend upon its diameter but upon the head and the amount of water applied to it. Where a very considerable amount of power is wanted under a comparatively low head, a wheel of larger diameter is necessary to admit of buckets of corresponding capacity, as also the application of two or more nozzles, for purpose of multiplying power. Where wheels of standard sizes do not meet the requirements of any particular case, special wheels can be made to suit the conditions presented. They are sometimes made as large as 15 to 20 and 25 feet diameter to conform to machinery to be operated, admitting in this way of direct connection to crank shafts of pumps, compressors, etc. When the requirements demand slow speed and a small amount of power under high heads, wheels of large diameter may be used with such buckets and nozzles as may be needed for the most economical effect.

The speed of the wheel being determined by the water head or pressure, the diameter can be made to conform to the speed required, the buckets and nozzles proportioned to amount of water available and power wanted, a special adaptation being thus made as far as possible in each case to the conditions under which the wheel is to work. The facility with which such adaptation can be made to all varying conditions, is one of the marked and distinguishing features of the Pelton Wheel, and admits of application to every possible service in the most simple, economical, and efficient way.

### USE OF TWO OR MORE NOZZLES.

When more than one nozzle is used, it is for the purpose of increasing the power by applying more water, or for securing a higher speed than can be obtained from a larger wheel with a single nozzle. By using multiple nozzles sufficient power can often be obtained from a wheel of small diameter to admit of direct connection to shaft of dynamo or other high-speed machinery without intermediate gearing, or of giving such increase of speed as to admit of belting direct without the use of counter-shafting and pulleys. Several illustrations are herein given of such application to compressors, blowers, pumps and other machinery, the advantages of which, from every point of view, are most apparent. Where two or more streams are applied to a wheel, some of them can be shut off when the additional power they afford is not wanted, or when for any reason the water supply fails in part—the reduced quantity not affecting in any appreciable degree either the percentage of efficiency or speed of wheel. The power in such case will depend upon the amount of water used, assuming that the head is always the same.

### CONDITIONS AS TO HEAD.

Experiments have shown that the Pelton Wheel will give as high an efficiency as any form of turbine under heads as low as 10 to 12 feet, but its construction does not admit of handling sufficient water to develop any considerable amount of power under so low a head within a reasonable limit of cost. It is not, therefore, recommended for heads of less than 30 to 40 feet except where but a small amount of power is required. If not more than 10 to 15 or 20 horse-power is wanted, it can be produced on a Pelton Wheel cheaper and more efficiently than by any other means under heads not higher than 15 to 20 feet. For heads above 40 or 50 feet it offers superior advantages under any conditions or for any class of service.

For what are termed high heads, say 100 feet and upwards, no other wheel is entitled to any consideration. As regards extreme pressure there is no practical limit to the head under which they may be safely and efficiently operated.

Upwards of 100 Pelton Wheels are now running under heads of more than 1,500 feet, and several under heads above 2,000 feet, all with a high degree of efficiency and with practically no expense in the way of repair.

## CONSERVATION OF HEAD.

As water in most cases where available for power has a positive commercial value, the most advantageous and profitable use of it should be considered. In this relation not only the wheel but the pipe which supplies the motive power has an important bearing upon the result. The loss of head by the friction of water in its passage through a line of pipe is more serious than is generally supposed, and must be carefully estimated in all cases where water is conveyed in this way any considerable distance.

Head, in this connection, is simply a convertible term for power, and has a direct relation to value. It should, therefore, be conserved to as great an extent as the circumstances of the case will justify. Where both the water supply and head are limited, the pipe should be of sufficient capacity to avoid as far as possible loss of head. Where water is abundant and a very considerable head can be obtained, a loss in this way may be justified to a larger extent to save cost in pipe.

The conditions presented must always determine what is most advisable, only considering the great advantage such a power affords and the fact that an increase beyond present needs is often wanted; it is well in planning a power station to make liberal provision in this regard.

### ADVANTAGES OF A HIGH HEAD.

The principle upon which the PELTON WHEEL operates being that of direct pressure, makes a high head desirable when it can be obtained at any reasonable cost, as the amount of water required to develop a given power decreases in direct ratio as the head or pressure increases. The additional length of pipe necessary in such case is often more than compensated for by its reduced size and weight, as also by the fact that the same results can be obtained with a smaller wheel and less water. The lessened expense in this way frequently justifies a largely increased expenditure in pipe line for the purpose of securing a higher head, especially as a small amount of water can often by such means be made most serviceable for electric lights or other purposes. The conditions presented in each particular case must, however, determine in regard to this.

### WIDE RANGE OF VARIATION.

One of the most notable features of the PELTON SYSTEM is the facility with which adaptation can be made to widely different conditions of water supply and power, without loss of useful effect.

This is accomplished by a simple change of nozzle tip, varying the size of the stream thrown on to the wheel—the power of which may be varied by this means from its maximum down to 25 per cent of same without appreciable loss—thus working to its full capacity with an ample water supply, or to the same relative advantage with a reduced quantity, when for any reason the supply fails in part.

The advantage of this means of adjustment to such varying conditions is apparent, as there are few cases where the water supply is not reduced at certain seasons of the year, making its economical and efficient use at such times most desirable. This arrangement also admits of using, without disadvantage, a wheel of larger capacity than present requirements demand, with reference to an increase of power when wanted. No other wheel admits of such changes without serious loss in useful effect.

Variations of construction as to diameter of wheel, size of buckets, number of streams applied, also admit of adaptation to all conditions and requirements of service either as to speed or power, in the most simple and efficient way, and at the smallest possible cost.

## DURABILITY AND RELIABILITY.

These features next to efficiency are conceded to be of the first importance in machinery of this character, for any purpose, more especially when operating electric plants—in which water power is now so important a factor. A comparison of the PELTON with other wheels will show its many advantages as regards simplicity of construction, absence of wearing parts, and small cost of maintenance, the latter amounting, in the majority of cases, to practically nothing for an indefinite time.

Many wheels may be referred to that have been running continuously for twelve and fifteen years under most unfavorable conditions without any expense whatever and without any impairment of efficiency. Water carrying sand and grit so destructive to other forms of wheels, has no effect upon this except under extreme high heads, when a set of buckets may be occasionally required, which involves small expense.

### FOR MILL AND MINE WORK.

The modern and most approved plan of operating machinery of this character, is that of having separate wheels for the various departments of work, such as batteries, crushers, concentrators, pumps, compressors, blowers, dynamos, etc. By this means the different classes of machinery are under separate control, making a governor unnecessary as well as in large part all intermediate gearing, the cost of the additional wheels for such a distribution of power being generally more than compensated for by the lessened expense of governor and connections referred to.

In locating wheels it is desirable to have them at the lowest point that will allow of free discharge within reasonable distance of the connecting machinery. Where the water supply is limited, the wheel running the crusher may be set high enough to use the discharge for the batteries.

### COMPARISON WITH TURBINES.

Under any reasonable head up to say 40 or 50 feet, with clear water free from sand and grit, it is admitted that turbines are ordinarily reliable and some of the best types efficient both at full and part gate, while with extremely low heads they are the only form of wheel that can be advantageously used. When operated under higher heads than above named—a practise not indorsed by the more conservative makers—it is well known that they are subject to serious wear and consequent loss of efficiency even when run under the most favorable conditions.

All mountainous regions, more especially in tropical countries, such as Mexico, Central and South America, are locations which present the most serious objections to the use of turbine wheels under any conditions of head. The streams furnishing power in such localities are subject to sudden freshets from excess of rainfall, and carry at such times grit and sand sufficient to destroy any turbine in a very short time. They also carry roots, leaves, and other trash that fill the vanes and choke the wheels to such an extent as often to prevent their running until the obstructions are removed involving a degree of unreliability that discredits the many advantages such a power ought always to afford.

The PELTON WHEEL, on the contrary, is wholly free from any such uncertainty, annoyance or trouble. The buckets, being opened, discharge freely anything that may be thrown into them. No instance has come to notice in the 10,000 wheels now running where complaint has been made from any such cause, or where any other wheel has ever been substituted for a PELTON. Large numbers of turbines have been replaced by PELTON wheels in the past few years, many running under heads not exceeding 30 to 50 feet, with such results as to more than compensate for the increased outlay.

The Quintex nozzle wheel, described and illustrated on page 26, is of special form and construction, intended particularly to meet such conditions as are here mentioned.

# MEANS OF UTILIZING WATER-POWERS.

The idea of a water-power is generally associated with a river or large stream, an expensive dam—huge flume—heavy grading and stone work—massive turbine wheels—pits—curbing and penstock, all involving so large an outlay as to make a power so produced of doubtful advantage, especially when so much is sacrificed in location of works as is generally necessary in such cases.

Not more than 5 per cent of the available water-powers in any part of the world, it is claimed, have thus far been utilized—a fact not to be wondered at, considering the crude and expensive methods that have so long prevailed. By means of the PELTON SYSTEM only a small diverting dam is required, then a pipe line running along the surface of the ground to the power station, which may be located at any convenient point, and high enough to be out of the reach of floods.

In this way a small trout brook will often furnish as much power as a large stream with a low head, in a much more convertible form, and at probably not more than one-fourth the outlay.

## NOTES IN REGARD TO MOUNTING WHEELS.

For mining purposes, wheels are generally mounted on timber framework, as illustrated on page 8. It is, in most cases, advisable to build this frame on the ground from detailed drawings furnished, rather than to pay freight on same for any considerable distance. With reasonable care in preparing foundations, this manner of setting is reliable and serviceable.

If suitable timber or lumber is not available, the frame is furnished when desired, in which case the wheel is erected on frame, all pieces marked, then taken down and packed for shipment. In this way the parts can readily be assembled on the ground.

For large and permanent plants, masonry or concrete foundations are advised. In many cases the wheels are mounted on iron bed plates, with iron housings and bracket stands for journal bearings with regulator attached—an especially compact and desirable arrangement for electrical work. Wheels so mounted are fitted with ring-oiling boxes, conforming, in both design and construction, to the best modern practise in this class of machinery.

Construction plans furnished with all wheels sent out, so that any competent mechanic can instal them without difficulty.

## MISCELLANEOUS NOTES.

The power of wheels listed is in all cases based upon effective heads, no allowance being made for friction loss. In ordering wheels the maximum power wanted should always be given; also full data as to water supply.

Different heads or falls can not be utilized in one pipe line nor on the same wheel. Two or more wheels can be run on the same shaft either for increase of power or speed, by using wheels of smaller diameter. Deflecting nozzles are often an advantage on wheels running under high heads both for purpose of regulation and relief to pipe line.

When the maximum power is wanted, or where economy of water is an object, the wheels should run with the gate wide open, using nozzle tips of such size as may be required to give the necessary power. When the water supply is diminished, the size of the nozzle tip should be reduced so as to maintain full head on the pipe line.

Where the wheel does not run in the right direction for the connecting machinery, it can be reversed and the water brought around to the wheel gate by a long-radius elbow, or it can be delivered on top of the wheel, the latter involving a loss of head equal to the diameter of wheel. In exceptional cases, to avoid cross belts, wheels can be run on vertical shafts.

By means of rope or sprocket chain attached to hand wheel of gate it can be operated at any convenient point even a considerable distance from the wheel.

STANDARD WHEEL MOUNTED ON WOOD FRAME, WITH HOUSING REMOVED.

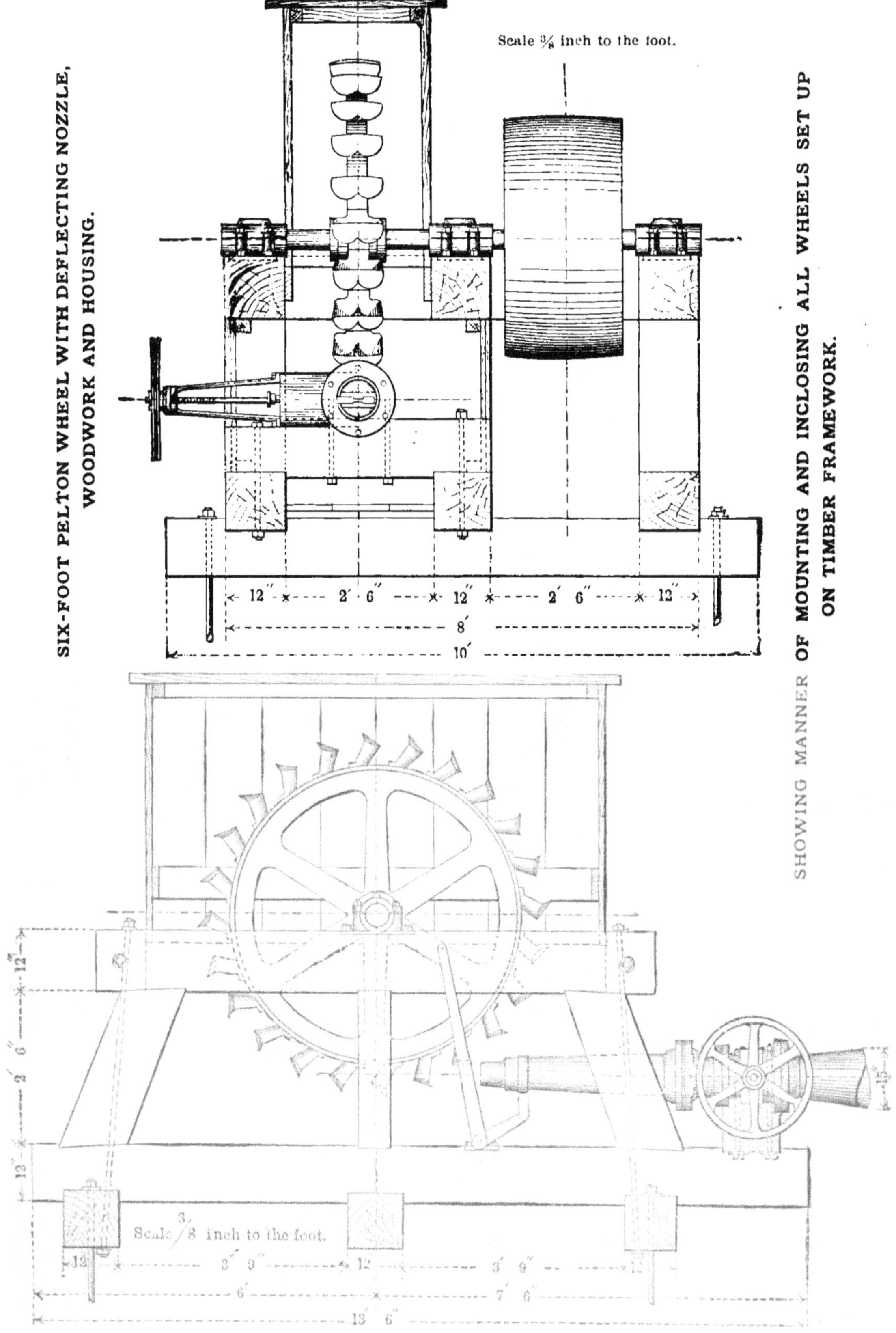


SIX-FOOT PELTON WHEEL WITH DEFLECTING NOZZLE, WOODWORK AND HOUSING.

SHOWING MANNER OF MOUNTING AND INCLOSING ALL WHEELS SET UP ON TIMBER FRAMEWORK.

PELTON WHEEL MOUNTED ON CHANNEL IRON FRAME.

The above print shows a standard double nozzle Pelton Wheel mounted on a channel iron frame. The housing may be of sheet iron or wood. The nozzles are fitted with cut-off hoods, operated by a hand wheel. An automatic governor can be attached when the variations of load are so frequent as to make such regulation necessary. Single nozzle wheels are mounted in the same manner.

This style of frame is substantial and durable and very desirable in many localities.

## PELTON WHEEL MOUNTED ON TIMBER FRAME.

The manner of mounting wheel, shown in print below, is generally used in mining districts, where timber is available. This wheel has also a double nozzle fitted with cut-off hoods, operated by a hand wheel. By means indicated the volume of water applied to the wheel may be reduced without any loss of head, affording a very convenient means of adapting the wheel to varying power requirements. An automatic governor can be attached when desired.

## MOTOR AND DYNAMO DIRECT CONNECTED.

The print on top of page 15 shows a Pelton Motor connected direct to shaft of a dynamo with an insulated coupling, both wheel and dynamo being mounted on a cast-iron bed plate, making a most simple and effective combination. The wheel in a plant of this character is made of a diameter to give proper speed to the dynamo under the water head available and of such capacity as may be desired, limited only by head and supply.

Where the water head does not admit of giving proper speed to dynamo by such direct connection, it can be run by belt without intermediate gearing. This application it is evident can be made to any style of dynamo.

Where any considerable head can be obtained, a small amount of water will afford a fine illuminating plant at a moderate cost, and one which, with free water, can be run without any expense whatever, thus affording great inducement to utilize all sources of water-power. Where reasonable water rates can be obtained such plants are often run from street mains at comparatively small cost.

# THE PELTON WATER MOTOR.

Believing that there was an imperative demand for a higher grade and more efficient WATER MOTOR for light power than had ever been put upon the market, THE PELTON WATER WHEEL CO., in response to this demand, were induced some years ago to get up a series of MOTORS inclosed in iron cases, embracing the general design and principle of their larger wheels, of a much more substantial and better character of mechanical work than was represented in any machine of the kind heretofore made. Though but a comparatively short time since these MOTORS were introduced in the present form, the demand has exceeded all expectations, showing an appreciation of their efforts to meet the want here referred to, and giving assurance that they will soon supersede all other water motors, as PELTON WHEELS have already done to a large extent in a wider range of service.

IT IS CONCEDED that nothing is so convenient, and, with any reasonable water rates, affords so cheap a power, as a water motor, especially for intermittent service. The fact that with the many advantages such motors offer, comparatively few are used, can only be accounted for by the high cost of power so obtained on account of the low efficiency and unserviceable character of the class of machines generally in use. Without disparagement of any particular motor it is only just to say that, notwithstanding the extravagant claims made for them,

**NO OTHER MOTOR NOW ON THE MARKET**

will bear critical inspection either as to mechanical detail or principle of operation. In fact, but few have the FIRST CLAIM to recognition as well designed and thoroughly serviceable machines.

THE PELTON WATER MOTOR is presented as being thoroughly scientific in principle, of excellent mechanical design, and possessing in the same relative degree the remarkable power of the PELTON WHEELS, which has given them a reputation for reliability, economy and efficiency that commands the trade wherever these advantages are understood and appreciated.

This motor is so superior to all others as regards strength, durability, economy of water, and in fact all that constitutes a HIGH-CLASS WATER MOTOR, as to admit of no comparison. As water for running motors is usually taken from street mains where, in many cases, high rates prevail, it is of the first importance that great economy and the best useful effect be secured. Where rates are not so high as to be absolutely prohibitory, the PELTON will be found to afford THE MOST ECONOMICAL, as well as THE MOST CONVENIENT, power that it is possible to obtain. It is warranted to do a given amount of work with from 25 to 50 per cent. less water than any other now on the market. Under all ordinary conditions this MOTOR will run for years without repair and with no attentionother than lubrication of journals. It is always in order, ready for instant service and

**GOOD FOR ANY DEMANDS THAT MAY BE MADE UPON IT**

within the limit of its capacity.

ATTENTION IS CALLED TO THE FACT that by a simple change of tips, varying the size of the stream thrown upon the wheel, a great variation of power can be obtained without loss of efficiency, a result only accomplished in other motors by throttling the water, which involves a material loss of pressure. This means of adapting the motors to varying quantities of water, gives them a wide range of utility and affords the greatest possible economy by using only the amount of water required for the work in hand.

The illustration of the 12-inch motor represents the form of the 6-inch and 15-inch as well. The 18-inch, 24-inch, and 30-inch being designed ordinarily for heavy duty, are of a somewhat different construction, being of the type illustrated by the 24-inch. The out-board bearing and base plate. shown in cut of the latter are not required nor furnished on standard motors under heads below 250 feet. The two largest motors are very heavy and substantial machines, and are capable of developing from 10 to 100 horse-power,

**DEPENDING UPON WATER AVAILABLE AND PRESSURE.**

PELTON MOTORS are especially adapted for running dynamos for isolated electric light plants. For this purpose, where the pressure admits and the power is not required for running other machinery, both wheel and dynamo can be mounted on a cast iron bed plate and the shafts connected by a coupling. Such means of connection, as well as the same applied to blowers for ventilation, is illustrated in this circular. Where the power is wanted for other use through the day, a belt connection with shifting pulley is used.

IT IS NEEDLESS TO SPECIFY THE GREAT VARIETY OF USES to which these motors are adapted, but the following list embraces some of the machinery to which they are applicable as power:—

Dynamos for Electric Lights,—Telephone Generators,—Passenger and Freight Elevators,—Printing Presses,—Wine Presses,— Power Pumps,—Woodworking Machinery,—General Machine Tools,—Exhaust Fans or Blowers for ventilating mines as well as public or private buildings,—Ice Cream Freezers,—Churns and various Dairy Machines,—Dental Lathes,—Carpenter Shops,—Sugar Mills,—Coffee and Spice Mills,—Coffee Roasters,—Meat Choppers,—Coffee Hulling Machinery,—Gas Machines,—Tobacco Machinery,—Grain Mills,—Hay Cutters,—Organs,—Sewing Machines, etc.

**All motors are provided with automatic ring oiling journals.**

## THE PELTON WATER MOTOR

### PRICE LIST AND WEIGHTS.

6 Inch Motor, $30, weight 30 lbs., pulley 3 inches diameter, ½ inch v groove.
12 Inch Motor, $60, weight 120 lbs., pulley 4 inches diameter x 4 inch face.
15 Inch Motor, $125, weight 220 lbs., pulley 5 inches diameter x 4½ inch face.
18 Inch Motor, $175, weight 350 lbs., pulley 6 inches diameter x 5 inch face.
24 Inch Motor, $275, weight 700 lbs., pulley 8 inches diameter x 8½ inch face.
30 Inch Motor, $350, weight 900 lbs., pulley 15 inches diameter x 10 inch face.

The above list covers MOTORS for all ordinary service up to a head of 25C feet or a pressure of 108 pounds. For heads in excess of this, heavier machines are furnished, for which special quotations are made. The prices are net cash, and include stop valve, driving pulley, and three interchangeable tips of different sizes to give such variation of power as may be desired.

COMPARISON IN PRICE can only be made with reference to the work accomplished, hence, taking into consideration the CAPACITY and EFFICIENCY of the PELTON MOTOR, it is proper to say that the prices above named are much lower than on those of any other make. The speed of these Motors is so uniform that Governors are only needed where a variety of machinery is run in an irregular way.

ATTENTION IS CALLED TO THE FACT that parties using water motors are often misled by the claims made as to efficiency, which in the majority of cases are greatly in excess of anything ever realized in practice, but which, without any means of disproving them or opportunity for comparison, go unchallenged. THE PELTON MOTOR is guaranteed to fully realize in practical working all claims made, provided it is properly installed in accordance with instructions.

### DIRECTIONS FOR SETTING AND RUNNING.

SPEED OF WHEEL: These Motors should in all cases be geared to run the number of revolutions given in tables; otherwise they will work at a great disadvantage, and will fail to develop their full power. The speed is determined by the ACTUAL PRESSURE when the wheel is running. To ascertain this, it is advisable to place a pressure gauge on the pipe near the inlet valve, and note both the standing and running pressure. Then gear the machinery it is proposed to run to the speed indicated in table under such running pressure. The difference between standing and running pressure determines the loss of head by friction in the pipe.

SIZE OF PIPE: Care should be taken that the supply pipe be of sufficient capacity to avoid undue loss of pressure by friction, especially when the most economical use of water is desired. The conditions of use vary so much that it is difficult to give any information in this regard that will apply to the majority of cases. The following will, however, indicate the smallest size that shou'd be used for a short line:—

6 Inch Motor . length not exceeding 100 feet . 1¼ inch diameter pipe.
12 Inch Motor . length not exceeding 100 feet . 2 inch diameter pipe.
15 Inch Motor . length not exceeding 100 feet . 2½ inch diameter pipe.
18 Inch Motor . length not exceeding 100 feet . 3 inch diameter pipe.
24 Inch Motor . length not exceeding 100 feet . 3½ inch diameter pipe.
30 Inch Motor . length not exceeding 100 feet . 5 inch diameter pipe.

For any lengths in excess of this, a proportionate increase in size should be made to cover friction. THE DISCHARGE PIPE SHOULD BE OF AMPLE CAPACITY, so that there may be no back water on the wheel, which would materially lessen its efficiency. Right-Angle Elbows or short turns SHOULD BE AVOIDED, as they increase the friction and lessen the pressure. The diameters of driving pulleys given are standard, but they can be varied to suit any required conditions as to speed. Rope sheaves can be substituted for pulleys, and are advised when the power is carried any considerable distance.

### THE MOTOR SHOULD BE LOCATED

at the lowest practicable point to avail of all the pressure obtainable. It is better to carry power by rope or belt some little distance than to sacrifice head to convenience of location. Where motors are run by water supplied from street mains, a liberal allowance should be made as regards capacity, on account of the varying pressure of such source of supply. When water is taken from mains that do not admit of a single tap of sufficient size to supply the service pipe, two or more taps can be made uniting a little distance from the main in a single pipe.

Three tips are sent with each machine, the largest of maximum capacity. It is advised to experiment with these, using the smallest first, and then the larger if necessary to give proper speed and power. The change of tips on the 6-inch, 12-inch, and 15-inch motors is made by taking off the valve, when the tip can be removed and another substituted. On the other sizes, this change is make by taking off the small plate on the lower left-hand side of the case and unscrewing the tip by means of a wrench. The pipe connection with gate should be made with a union.

SPECIAL ATTENTION is called to the following points, which SHOULD BE CAREFULLY OBSERVED in setting Motor: THAT THE SUPPLY PIPE IS OF AMPLE CAPACITY TO AVOID UNDUE LOSS OF HEAD BY FRICTION; THAT THE WHEEL HAS A FREE DISCHARGE, AND THAT IT IS GIVEN THE SPEED DUE TO PRESSURE AS SHOWN IN TABLES.

PARTIES WRITING FOR INFORMATION SHOULD STATE FULL PARTICULARS AS TO SOURCE OF WATER SUPPLY. If from street main, the average working pressure; if from spring or stream, the amount of water available, with vertical head or fall, length of pipe necessary to convey it to Motor; also a description of the machinery to be run, power required, size and speed of pulleys to which it is proposed to belt.

12-INCH MOTOR.

MOTOR AND DYNAMO DIRECT-CONNECTED.

## COMPARISONS OF VARIOUS WATER MOTORS.

MADE IN THE MECHANICAL DEPARTMENT OF THE UNIVERSITY OF MICHIGAN
UNDER THE DIRECTION OF PROF. M. E. COOLEY.

| NAME OF MOTORS. | CHICAGO. | BACKUS. | PELTON. |
|---|---|---|---|
| Catalogue Number or Size | No. 12. | 22 inch. | No. 2, 12 inch. |
| Size of Nozzle | 1/4" to 3/8" | 1/4" to 3/8" | 1/4" to 1/2" |
| Style of Nozzle | Conical, with Cylindrical Tip. | Conical, with Cylindrical Tip. | Conical, with Cylindrical Tip. |
| List Price, with Pulley | $125.00 | $85.00 | $60.00 |
| Rated Horse Power | 50 to 100 lbs. pressure 2 to 5 | 75 to 150 lbs. pressure 3 to 4½ | At 100 lbs. 4.9 |

### Comparison with 3-8 inch Nozzle under 100 lbs. Pressure.

| | | | |
|---|---|---|---|
| Theoretical Spouting Velocity in Feet per Second | 122 | 122 | 122 |
| Theoretical Discharge in Cubic Feet per Second | 0.0935 | 0.0935 | 0.0935 |
| Theoretical Discharge in lbs. per Second | 5.83 | 5.83 | 5.83 |
| Theoretical Horse Power | 2.448 | 2.448 | 2.448 |
| Co-efficient of Discharge | 0.97 | 0.97 | 0.97 |
| Actual Discharge in lbs. per Second | 5.655 | 5.655 | 5.655 |
| Efficiency, per cent | 55 | 55 | 80 |
| Actual Horse Power | 1.306 | 1.306 | 1.900 |
| Relative Horse Power | 0.687 | 0.687 | 1.000 |
| Cubic Feet of Water per Minute | 5.44 | 5.44 | 5.44 |
| Cubic Feet of Water per Horse Power per Hour | 249.96 | 249.96 | 171.84 |
| Gallons Water per Horse Power per Hour | 1874.7 | 1874.7 | 1288.8 |
| Cost per Horse Power per Hour at 10c. per 1000 Gals | $0.187 | $0.187 | $0.129 |
| Relative Cost to Operate | 1.454 | 1.454 | 1.000 |
| Cost per Horse Power to Buy | 96.10 | 65.40 | 31.58 |
| Relative Cost to Buy | 3.04 | 2.07 | 1.00 |

MICHIGAN UNIVERSITY, ANN ARBOR, MICHIGAN, MAY 23RD, 1891. M. E. COOLEY.

**NOTE.**—The motors referred to in the first two columns are generally recognized as fairly representing the various other types on the market; in fact, they are claimed by the Manufacturers to be superior; it is not too much, therefore, to assume that they make a fair basis of comparison.

By reference to results it will be seen that the Pelton motor shows 45% higher efficiency than either of the others, hence it will cost only 55%—but a little more than one-half—as much to operate; while the cost of the motor per horse-power to buy is only one-half that of the Backus, and one-third that of the Chicago. All claims made for the Pelton are thus fully substantiated by the highest scientific authorities, indicating clearly that no other water motor is entitled to any consideration by purchasers.

In view of the great advantage the Pelton motor offers as shown by the above tests, as well as by results in practical working, many of the Water Companies in the larger cities have made such a discrimination in rates in favor of the Pelton as to practically exclude all others. This it may be said will be the case in all localities where economy of water is a consideration, with reference to making the most of such water supply as may be available for power.

The experiments above recorded it may be said were entirely disinterested, and made wholly from a scientific standpoint, while the high character of the Institution mentioned, the completeness of its equipment and thoroughness of its work—especially in the department of Dynamic Engineering—affords absolute assurance of accuracy in all details as well as conclusions.

## DUPLEX PUMP RUN BY A PELTON WHEEL.

The above cut shows a duplex power pump driven by a PELTON WHEEL attached to shaft of pump direct. Where the conditions do not admit of such connection, the wheel may be independent with belt or rope transmission. The following table furnishes data of some of the ordinary sizes for general service:

### DATA AND DIMENSIONS OF THE VARIOUS SIZES.

| Water Cylinder. | Stroke. | Gallons per Revolution, both Cylinders. | *Capacity per Minute at Moderate Speed, Gallons. | Suction Pipe. | Delivery Pipe. | Water Cylinder. | Stroke. | Gallons per Revolution, both Cylinders. | *Capacity per Minute at Moderate Speed, Gallons. | Suction Pipe. | Delivery Pipe. |
|---|---|---|---|---|---|---|---|---|---|---|---|
| 4 | 6 | 1.28 | 102 | 3 | 2 | 12 | 24 | 46.97 | 940 | 12 | 10 |
| 6 | 8 | 3.91 | 234 | 5 | 4 | 14 | 18 | 48.00 | 1280 | 14 | 12 |
| 7 | 10 | 6.66 | 319 | 6 | 5 | 14 | 24 | 64.00 | 1280 | 14 | 12 |
| 8 | 12 | 10.44 | 418 | 8 | 6 | 16 | 18 | 62.64 | 1670 | 16 | 14 |
| 9 | 12 | 13.20 | 528 | 8 | 6 | 16 | 24 | 83.52 | 1670 | 16 | 14 |
| 10 | 12 | 16.32 | 653 | 10 | 8 | 18 | 24 | 105.76 | 2115 | 18 | 16" |
| 10 | 18 | 24.48 | 653 | 10 | 8 | 18 | 36 | 158.64 | 2115 | 18 | 16 |
| 12 | 12 | 23.48 | 940 | 12 | 10 | 20 | 24 | 130.52 | 2610 | 20 | 18 |
| 12 | 18 | 35.20 | 940 | 12 | 10 | 20 | 36 | 195.80 | 2610 | 20 | 18 |

* The capacity given is based upon a piston speed of 80 feet per minute, which can be increased if desired.

Special pumps with suitable wheel connections made of any capacity and for any required duty. Cornish Plunger Pumps may also be run in the same way, the connecting rod being attached to crank-pin on wheel, the latter being made of size to give proper speed to pump—a similar application to that shown in connection with air compressors

NOTE.—A pumping plant of same construction as shown in above cut—capacity 50,000 gallons per hour—was constructed 5 years ago for one of the largest Central American cities, a three-nozzle PELTON WHEEL, running under 60-foot head, furnishing the power. The plant has been in continuous operation during the time named, giving most economical, reliable and efficient service.

## DATA REQUIRED FOR ESTIMATES.

CORRESPONDENTS APPLYING FOR INFORMATION SHOULD GIVE THE FOLLOWING DATA OR AS MUCH THEREOF AS APPLIES TO THEIR CASE :

FIRST. Amount of water available, either in miner's inches or flow *per minute* in gallons or cubic feet. If in miner's inches, give the size of the aperture in the measuring box and the head above the center of the same. When the water supply varies at different seasons, give the smallest and largest amount, with a general average, and state if it is desired to have the Wheel of capacity to utilize the full amount.

SECOND. Head or vertical fall of water from ditch, flume or other source of supply to the point where Wheel is to be located, with accuracy if possible; otherwise approximately.

THIRD. Length and diameter of pipe line conveying water to Wheel. If not already laid, consult tables as to proper size or write for information, giving full particulars as to length, and such other data as is called for herein.

FOURTH. Horse power required and what it is designed to run. If for quartz mill, give the number and weight of stamps, with lift and drop per minute. Also describe in detail, any other machinery it is proposed to run, with speed, etc. If for dynamos, give capacity and speed and whether for power or light, distance of transmission, etc.

FIFTH. For pumping give amount of water to be raised and vertical depth. For hoisting give weight of cage, car and cable, as also load it is desired to carry, together with the vertical height, speed, etc.

SIXTH. Where the size of flume or ditch is given showing volume of water, give in all cases the velocity of the current, or the grade on which the flume is laid. Means of ascertaining this will be found in explanations accompanying tables.

SEVENTH. Do not give as water supply an amount that will fill a certain sized pipe, calculations based upon such information being difficult and unreliable.

EIGHTH. Where reference is made to information contained in catalogue, state the tables from which such information is obtained, and the edition of the catalogue.

NINTH. With order, the fall or effective head, as also the horse power the Wheel is intended to develop, should be given with as much accuracy as possible. Wheels in no cases furnished without gate and nozzle, these being necessary to make a proper application of the stream to the Wheel.

TENTH. Orders for Wheels or Motors from other than well known firms of established character, unless accompanied by remittance, should be sent through some responsible commission house, or available reference given. Address, destination and route should also be plainly stated.

HOME CORRESPONDENTS will please write plainly address, giving post-office with County and State or Territory.

FOREIGN CORRESPONDENTS will please give plainly address, name of post-office, with country, colony or province, and such other information as may be desirable to insure safe and prompt transmission.

## TERMS AND OTHER INFORMATION.

Prices quoted in preceding tables are net and include wheel, shaft, boxes, collars, gate, and nozzle with interchangeable nozzle tips to give such variation of power as may be desired.

TERMS CASH in current funds. One-third with order, balance on shipment, or subject to sight draft with Bill-Lading, or whole amount may be subject to draft against documents, with proper reference.

While not responsible for either safe or prompt delivery after shipment in good condition, we will at all times cheerfully cooperate with our patrons in having their claims equitably adjusted.

Driving pulleys for wheel shaft, of any required size, as also timber frame for mounting wheel, with frame rods and bolts, are an extra charge, but furnished when wanted, at reasonable rates.

The weights of the 3-foot and 4-foot wheels come within the limit for mule packing. The larger sizes for this requirement are made in sections not exceeding 250 to 300 pounds. In such case 12 per cent must be added to list price.

Preliminary plans regarding any proposed work submitted, upon sufficient data being given. Construction plans embracing full details furnished with all work sent out.

Parties having water-powers they are desirous of utilizing will be furnished with reliable information as to the best means of making them available for purpose intended. Applications of this character should give the fullest particulars as to location and all surrounding conditions.

These, embraced in general terms, are—amount of water available, highest practicable head, power required, class of machinery to be run, approximate length of pipe line from source of supply to wheel station, as well as any other details that may assist in giving an intelligent idea of the situation. A diagram of pipe line showing pressure at various points as well as location of power station is also desirable.

Next to the proper adaptation of wheels to conditions of service, the most important thing in a water-power plant is the pipe line and the various connections thereto. If not properly proportioned with reference to size, strength, bends, and connections, the efficiency will be greatly impaired and much trouble will result. Cases are constantly coming to notice where serious loss and disappointment have been occasioned by faulty arrangement and construction, requiring large expenditures to remedy mistakes in this way which should have been avoided.

An experience of more than fifteen years in planning and executing works of this character, with unvarying success, will afford assurance that such information as may be given by this company, regarding the best means of utilizing water-powers, is entitled to the fullest confidence; also that all work furnished will be adapted to the situation and requirements of the case, while involving all the elements of security, economy, and reliability that should attach to work of so important a character.

Where there is a substantial compliance with conditions given as regards water supply and head, there can be no disappointment in results, as these can be calculated upon with all the certainty that applies to any problem in mathematics.

Attention is called to our unusual facilities for supplying at the lowest ruling rates riveted SHEET STEEL or LAP-WELD PIPE, together with RELIEF VALVES, WATER GATES, RECEIVERS and other connections, as also SHAFTINGS, PULLEYS, COUPLINGS, JOURNAL BEARINGS, ROPE DRIVES, and other POWER-TRANSMITTING MACHINERY. Propositions submitted for entire power plants, of any capacity and under any conditions of service.

## PRICE LIST, POWER AND WEIGHT PELTON WATER WHEELS.

### STANDARD SIZES.

| Head in Feet. | 3 Foot Wheel. | | 4 Foot Wheel. | | 5 Foot Wheel. | | 6 Foot Wheel. | |
|---|---|---|---|---|---|---|---|---|
| | H. P. | Price. | H. P. | Price. | H. P. | Price. | H. P. | Price. |
| 20 | 1.50 | $220 | 2.64 | $285 | 4.18 | $350 | 6.00 | $400 |
| 50 | 5.98 | 220 | 10.60 | 285 | 16.63 | 350 | 23.93 | 400 |
| 80 | 12.04 | 230 | 21.44 | 295 | 33.54 | 370 | 48.16 | 425 |
| 100 | 16.84 | 240 | 29.93 | 300 | 46.85 | 380 | 67.36 | 450 |
| 150 | 31.01 | 250 | 55.08 | 320 | 86.22 | 400 | 124.04 | 475 |
| 200 | 47.75 | 265 | 84.81 | 350 | 132.70 | 425 | 191.00 | 500 |
| 250 | 66.74 | 280 | 118.54 | 375 | 185.47 | 450 | 266.96 | 550 |
| 300 | 87.73 | 300 | 155.83 | 400 | 243.82 | 485 | 350.94 | 625 |
| 350 | 110.56 | 320 | 196.38 | 425 | 307.25 | 550 | 442.27 | 700 |
| 400 | 135.08 | 340 | 239.94 | 460 | 375.40 | 625 | 540.35 | 800 |
| 450 | 161.19 | 360 | 286.31 | 500 | 447.95 | 700 | 644.78 | 900 |
| 500 | 188.80 | 390 | 335.34 | 540 | 524.66 | 775 | 755.20 | 1000 |
| 550 | 223.76 | 425 | 397.43 | 600 | 621.82 | | 895.04 | |
| 600 | 248.16 | 450 | 440.77 | 675 | 689.63 | | 992.65 | |
| 700 | 312.73 | 500 | 555.46 | 750 | 869.66 | | 1250.92 | |
| 800 | 382.69 | | 678.66 | | 1061.81 | | 1528.36 | |
| 900 | 455.91 | | 809.82 | | 1267.02 | | 1823.76 | |
| 1000 | 534.01 | | 948.48 | | 1483.97 | | 2136.04 | |
| | Weight, 700 to 1000 lbs. | | Weight, 1000 to 1700 lbs. | | Weight, 1400 to 2100 lbs. | | Weight, 2100 to 3000 lbs. | |

## PRICE LIST PELTON WATER MOTORS.

| Head in Feet | 6 Inch. | | 12 Inch. | | 15 Inch. | | 18 Inch. | | 24 Inch. | |
|---|---|---|---|---|---|---|---|---|---|---|
| | H. P. | Price. | H. P. | Price. | H. P. | Price. | H. P. | Price. | H. P. | Price. |
| 20 | .05 | $30 | .12 | $60 | .20 | $125 | .37 | $175 | .63 | $275 |
| 50 | .21 | 30 | .49 | 60 | .84 | 125 | 1.49 | 175 | 2.65 | 275 |
| 80 | .43 | 30 | 1.00 | 60 | 1.70 | 125 | 3.10 | 175 | 5.36 | 275 |
| 100 | .60 | 30 | 1.40 | 60 | 2.32 | 125 | 4.21 | 175 | 7.49 | 275 |
| 150 | 1.10 | 30 | 2.58 | 60 | 4.37 | 125 | 7.75 | 175 | 13.77 | 275 |
| 200 | 1.70 | 30 | 3.97 | 60 | 6.74 | 125 | 11.93 | 175 | 21.20 | 275 |
| 250 | 2.38 | 30 | 5.56 | 60 | 9.42 | 125 | 16.63 | 175 | 29.63 | 275 |
| 300 | 3.13 | 35 | 7.31 | 70 | 12.38 | 140 | 21.93 | 200 | 38.95 | 300 |
| 350 | 3.94 | 35 | 9.21 | 70 | 15.61 | 140 | 27.64 | 200 | 49.09 | 300 |
| 400 | 4.82 | 40 | 11.25 | 80 | 19.07 | 150 | 33.77 | 220 | 59.98 | 320 |
| 450 | 5.75 | 40 | 13.43 | 80 | 22.76 | 150 | 40.29 | 220 | 71.57 | 320 |
| 500 | 6.74 | 50 | 15.73 | 90 | 26.66 | 160 | 47.20 | 230 | 83.83 | 330 |
| 600 | 8.75 | 50 | 24.26 | 100 | 34.90 | 160 | 62.04 | 230 | 110.19 | 330 |
| 700 | 11.02 | 60 | 30.57 | 110 | 43.99 | 175 | 78 18 | 250 | 138.86 | 360 |
| 800 | 13.47 | 70 | 37.35 | 120 | 53.73 | 190 | 95.52 | 275 | 169.66 | 390 |
| 900 | 16.07 | 80 | 44.57 | 130 | 64.12 | 210 | 113.98 | 300 | 202.45 | 420 |
| 1000 | 18.82 | 90 | 52.20 | 140 | 75.11 | 225 | 133.50 | 325 | 237.12 | 460 |

## PRICE LIST AND POWER TWO-NOZZLE WHEELS.

| Head in Feet. | 3-Foot Wheel. | | 4-Foot Wheel. | | 5-Foot Wheel. | | 6-Foot Wheel. | |
|---|---|---|---|---|---|---|---|---|
| | H. P. | Price. | H. P. | Price. | H. P. | Price. | H. P. | Price. |
| 20 | 3.00 | $300 | 5.28 | $375 | 8.36 | $ 425 | 12.00 | $ 500 |
| 30 | 5.52 | 300 | 9.76 | 375 | 15.38 | 425 | 22.08 | 500 |
| 40 | 8.48 | 310 | 15.16 | 380 | 23.70 | 435 | 33.92 | 510 |
| 50 | 11.96 | 325 | 21.28 | 385 | 33.26 | 440 | 47.86 | 520 |
| 60 | 15.68 | 340 | 27.88 | 390 | 43.54 | 450 | 62.72 | 530 |
| 70 | 19.76 | 360 | 35.16 | 400 | 55.02 | 460 | 79.04 | 550 |
| 80 | 24.08 | 380 | 42.88 | 420 | 67.08 | 475 | 96.32 | 575 |
| 100 | 33.68 | 400 | 59.86 | 450 | 93.70 | 500 | 134.72 | 600 |
| 150 | 62.02 | 440 | 110.16 | 500 | 172.44 | 575 | 248.08 | 700 |
| 200 | 95.50 | 480 | 169.62 | 575 | 265.40 | 650 | 382.00 | 800 |
| 250 | 133.48 | 525 | 237.08 | 650 | 370.94 | 750 | 533.92 | 900 |
| 300 | 175.46 | 550 | 311.66 | 680 | 487.64 | 800 | 701.88 | 960 |
| 350 | 221.12 | 575 | 392.76 | 710 | 614.50 | 860 | 884 54 | 1025 |
| 400 | 270.16 | 600 | 479.88 | 740 | 750.80 | 925 | 1080.70 | 1125 |
| 450 | 322.38 | 625 | 572.62 | 775 | 895.90 | 1000 | 1289.56 | 1225 |
| 500 | 377.60 | 650 | 670.68 | 800 | 1049.32 | 1075 | 1510.40 | 1350 |
| | Weight 950 to 1250 lbs. | | Weight 1250 to 2000 lbs. | | Weight 1900 to 2400 lbs. | | Weight 2400 to 4000 lbs. | |

NOTE.—Prices on all special wheels quoted on application, also on wheels of larger diameter or capacity than above given.

## EXPLANATION PELTON WATER WHEEL TABLES.

The tables on the following pages are prepared specially for the Pelton Wheel, and are not adapted to any other. The spouting velocity given is that due to the effective head, without any allowance for friction in either pipe line or nozzle. The peripheral velocity of the wheel should be one-half the spouting velocity of the water. Special wheels of intermediate diameters from those designated in tables are furnished when required, to meet any conditions as to speed, also as to capacity. In selecting a wheel for a given power, reference should, therefore, be had to the speed at which it runs, taking into consideration the speed of the driven machinery. In this way intermediate connections can often be dispensed with, affording a material saving in first cost, as well as maintenance and power.

## AUTOMATIC WATER WHEEL REGULATION.

Close and sensitive automatic regulation is conceded to be not only of the first importance but absolutely essential for safety and success in operating an electrical plant for either power or light. The lack of such regulation, it is well known, has been the serious objection to the use of water power for electrical purposes. This want has, however, now been supplied in so complete and satisfactory a way as to obviate all such objection.

Several types of governors are used by this company, no single one being adapted to all the varying conditions met with. Propositions for the regulation of electric plants can only be submitted after a full understanding of these conditions.

It is sufficient in this connection to say that we are prepared to guarantee that governors furnished will afford sensitive and reliable control, such as may be entirely satisfactory to any of the electrical companies with whose generators the wheels may be connected. No difficulty in this regard has been experienced in the large number of electrical installations made by this company subsequently referred to.

# PELTON WATER WHEEL TABLES.

The calculations for power in these tables are based upon the application of one stream to the Wheel, as also upon an 85 per cent efficiency and effective heads, no allowance being made for loss of head by friction in pipe. The smaller figures under those denoting the various heads give the equivalent pressure in pounds and spouting velocity of the water in feet per minute. The cubic feet measurement is also based on the flow per minute.

| HEAD IN FEET | SIZE OF WHEELS. | 6 Inch | 12 Inch | 15 Inch | 18 Inch | 24 Inch | 3 Foot | 4 Foot | 5 Foot | 6 Foot |
|---|---|---|---|---|---|---|---|---|---|---|
| **20** | Horse Power...... | .05 | .12 | .20 | .37 | .66 | 1.50 | 2.64 | 4.18 | 6.00 |
| 8 Lbs. | Cubic Feet........ | 1.67 | 8.91 | 6.62 | 11.72 | 20.83 | 46.93 | 83.32 | 130 36 | 187.72 |
| | Miner's Inches.. | 1.04 | 2.44 | 4.00 | 7.82 | 13.88 | 31.28 | 55.52 | 86.90 | 125.12 |
| 2151 97 | Revolutions........ | 684 | 342 | 274 | 228 | 171 | 114 | 85 | 70 | 57 |
| **30** | Horse Power...... | 10 | .23 | .38 | .69 | 1.22 | 2.76 | 4.88 | 7.69 | 11.04 |
| 13 Lbs. | Cubic Feet........ | 2.05 | 4.79 | 8.11 | 14.36 | 25.51 | 57.44 | 102.04 | 159.66 | 229.76 |
| | Miner's Inches... | 1.28 | 2.99 | 5.06 | 9.57 | 17.00 | 88.28 | 68.00 | 106.44 | 153.12 |
| 2635.62 | Revolutions........ | 837 | 418 | 335 | 279 | 209 | 139 | 104 | 83 | 69 |
| **40** | Horse Power...... | .15 | .35 | .59 | 1.06 | 1.89 | 4.24 | 7.58 | 11.85 | 16.96 |
| 17 Lbs. | Cubic Feet........ | 2.37 | 5 53 | 9 37 | 16.59 | 29.46 | 66.36 | 107.84 | 184.86 | 265.44 |
| | Miner's Inches... | 1.48 | 8.45 | 5.85 | 11.06 | 19.64 | 44.24 | 78.56 | 122.91 | 176.96 |
| 3043.39 | Revolutions........ | 969 | 484 | 887 | 323 | 242 | 161 | 121 | 96 | 80 |
| **50** | Horse Power...... | .21 | .49 | .84 | 1.49 | 2.65 | 5.98 | 10.60 | 16.63 | 23.93 |
| 21 Lbs. | Cubic Feet........ | 2 64 | 6.18 | 10.47 | 18.54 | 32.93 | 74.17 | 131.72 | 206.13 | 296.70 |
| | Miner's Inches... | 1.65 | 3.86 | 6.54 | 12.86 | 21.95 | 49.45 | 87.80 | 137.42 | 197.80 |
| 3402.61 | Revolutions........ | 1083 | 541 | 433 | 361 | 270 | 180 | 135 | 108 | 90 |
| **60** | Horse Power...... | .28 | .65 | 1.10 | 1.96 | 3.48 | 7.84 | 13.94 | 21.77 | 31.36 |
| 26 Lbs. | Cubic Feet........ | 2.90 | 6.77 | 11.47 | 20.31 | 86.08 | 81.25 | 144.32 | 225.80 | 325.00 |
| | Miner's Inches... | 1.81 | 4.23 | 7.16 | 13.54 | 24.05 | 54.16 | 96.20 | 150.52 | 216.64 |
| 3727.37 | Revolutions........ | 1185 | 592 | 473 | 395 | 296 | 197 | 148 | 118 | 98 |
| **70** | Horse Power...... | .35 | .82 | 1.39 | 2.47 | 4.39 | 9.88 | 17.58 | 27.51 | 39.52 |
| 30 Lbs. | Cubic Feet........ | 3.13 | 7.31 | 12.39 | 21.94 | 38.97 | 87.76 | 155.88 | 243.89 | 851.04 |
| | Miner's Inches... | 1.95 | 4 56 | 7.74 | 14.63 | 25.98 | 58.52 | 108.92 | 162.60 | 234.08 |
| 4026.00 | Revolutions........ | 1281 | 640 | 512 | 427 | 320 | 213 | 160 | 130 | 106 |
| **80** | Horse Power...... | .43 | 1.00 | 1.70 | 3.01 | 5.36 | 12.04 | 21.44 | 33.54 | 48.16 |
| 34 Lbs. | Cubic Feet........ | 3.35 | 7.82 | 13.25 | 23.46 | 41.66 | 93.84 | 166.64 | 260.73 | 875.36 |
| | Miner's Inches... | 2.09 | 4.88 | 8.28 | 15.64 | 27.77 | 62.56 | 111.08 | 173.82 | 250.24 |
| 4303.99 | Revolutions..... .. | 1368 | 684 | 546 | 456 | 342 | 228 | 171 | 137 | 114 |
| **90** | Horse Power...... | .51 | 1.20 | 2.03 | 3.60 | 6.39 | 14.40 | 25.59 | 40.04 | 57.60 |
| 39 Lbs. | Cubic Feet........ | 3.55 | 8.29 | 14.05 | 24.88 | 44.19 | 99.52 | 176 75 | 276.55 | 398.08 |
| | Miner's Inches... | 2.22 | 5.18 | 8.78 | 16.58 | 29 46 | 66.32 | 117.83 | 184.36 | 265.28 |
| 4565.04 | Revolutions........ | 1452 | 726 | 581 | 484 | 363 | 242 | 181 | 145 | 121 |
| **100** | Horse Power...... | .60 | 1.40 | 2.32 | 4.21 | 7.49 | 16.84 | 29.93 | 46.85 | 67.36 |
| 43 Lbs. | Cubic Feet........ | 8 74 | 8.74 | 14.81 | 26.22 | 46.58 | 104.88 | 186.32 | 291.51 | 419.52 |
| | Miner's Inches... | 2.33 | 5.46 | 9.25 | 17.48 | 31.05 | 69.93 | 124.21 | 194.34 | 279.72 |
| 4812.00 | Revolutions........ | 1530 | 765 | 612 | 510 | 382 | 255 | 191 | 152 | 127 |
| **110** | Horse Power...... | .69 | 1.62 | 2.74 | 4.86 | 8.64 | 19.44 | 34.58 | 54.11 | 77.76 |
| 48 Lbs. | Cubic Feet........ | 3.92 | 9.16 | 15.53 | 27.50 | 48.85 | 110.00 | 195.41 | 305.73 | 440.00 |
| | Miner's Inches... | 2.45 | 5.72 | 9.70 | 18.33 | 32.56 | 73.33 | 130.27 | 203.82 | 293.32 |
| 5016.87 | Revolutions........ | 1605 | 802 | 642 | 535 | 401 | 267 | 200 | 160 | 133 |
| **120** | Horse Power...... | .79 | 1.84 | 3.12 | 5.54 | 9.85 | 22.18 | 39.41 | 61.66 | 88.75 |
| 52 Lbs. | Cubic Feet........ | 4.10 | 9.57 | 16.21 | 28.72 | 51.02 | 114.91 | 204.10 | 319.33 | 459.64 |
| | Miner's Inches... | 2.56 | 5.98 | 10.13 | 19.15 | 34.01 | 76.60 | 136.06 | 212.89 | 806.43 |
| 5271.30 | Revolutions...... | 1677 | 838 | 671 | 559 | 419 | 279 | 209 | 167 | 139 |
| **130** | Horse Power...... | .89 | 2.08 | 3.53 | 6.25 | 11.11 | 25.02 | 44.46 | 69.53 | 100.08 |
| 56 Lbs. | Cubic Feet........ | 4.27 | 9.96 | 16.89 | 29.90 | 53.10 | 119.60 | 212.43 | 832.37 | 478.41 |
| | Miner's Inches... | 2.66 | 6.22 | 10.55 | 19.93 | 35.40 | 79.73 | 141.62 | 221.58 | 318.94 |
| 5496.54 | Revolutions........ | 1746 | 873 | 698 | 582 | 436 | 291 | 218 | 174 | 145 |
| **140** | Horse Power...... | .99 | 2.33 | 3.94 | 6.99 | 12.41 | 27.96 | 49.64 | 77.71 | 111.85 |
| 60 Lbs. | Cubic Feet........ | 4.43 | 10.34 | 17.53 | 31.03 | 55.11 | 124.12 | 220.44 | 344.92 | 496.48 |
| | Miner's Inches... | 2.76 | 6.46 | 10.95 | 20.68 | 36.74 | 82.72 | 146.96 | 229.94 | 330.88 |
| 5693.65 | Revolutions........ | 1812 | 906 | 725 | 604 | 453 | 302 | 226 | 181 | 151 |
| **150** | Horse Power...... | 1.10 | 2.58 | 4.37 | 7.75 | 13.77 | 31.01 | 55.08 | 86.22 | 124.04 |
| 65 Lbs. | Cubic Feet........ | 4.55 | 10.70 | 18.14 | 32.11 | 57.04 | 128.47 | 228.19 | 357.02 | 513.90 |
| | Miner's Inches... | 2.84 | 6.68 | 11.33 | 21.41 | 38.03 | 85.64 | 152.12 | 238.05 | 342.59 |
| 5893.41 | Revolutions........ | 1875 | 937 | 750 | 625 | 468 | 312 | 234 | 187 | 156 |
| **160** | Horse Power...... | 1.22 | 2.84 | 4.82 | 8.54 | 15.17 | 34.16 | 60.68 | 94.94 | 136.65 |
| 69 Lbs. | Cubic Feet........ | 4.73 | 11.05 | 18.74 | 33.17 | 58.92 | 132.68 | 235.68 | 368.73 | 530.75 |
| | Miner's Inches... | 2.95 | 6.90 | 11.71 | 22.11 | 39.28 | 88.46 | 157.12 | 245.82 | 353.84 |
| 6086.71 | Revolutions........ | 1938 | 969 | 775 | 646 | 484 | 323 | 242 | 193 | 161 |

# PELTON WATER WHEEL TABLES.

The calculations for power in these tables are based upon the application of one stream to the Wheel, as also upon an 85 per cent efficiency and effective heads, no allowance being made for loss of head by friction in pipe. The smaller figures under those denoting the various heads give the equivalent pressure in pounds and spouting velocity of the water in feet per minute. The cubic feet measurement is also based on the flow per minute.

| HEAD IN FEET | SIZE OF WHEELS. | 6 Inch | 12 Inch | 15 Inch | 18 Inch | 24 Inch | 3 Foot | 4 Foot | 5 Foot | 6 Foot |
|---|---|---|---|---|---|---|---|---|---|---|
| **170** | Horse Power...... | 1,38 | 3,11 | 5,28 | 9,35 | 16,61 | 87.42 | 66.46 | 103,99 | 149,68 |
| | Cubic Feet........ | 4.88 | 11,39 | 19.31 | 34.19 | 60 73 | 136.77 | 242.93 | 380.08 | 547.08 |
| 74 Lbs. | Miner's Inches.. | 3.05 | 7.11 | 12.06 | 22.79 | 40.48 | 91.18 | 161.95 | 253.38 | 864.72 |
| 6274.07 | Revolutions........ | 1995 | 997 | 798 | 665 | 498 | 332 | 249 | 199 | 166 |
| **180** | Horse Power...... | 1,45 | 3.39 | 5.75 | 10.19 | 18.10 | 40.77 | 72,41 | 113.30 | 163.08 |
| | Cubic Feet........ | 5.02 | 11.72 | 19.87 | 35.18 | 62.49 | 140.74 | 249.97 | 891.10 | 562.96 |
| 78 Lbs. | Miner's Inches... | 3.13 | 7.32 | 12.41 | 28.45 | 41,66 | 93.82 | 166.64 | 260.73 | 375.29 |
| 6455.97 | Revolutions........ | 2049 | 1024 | 820 | 683 | 513 | 342 | 256 | 206 | 171 |
| **190** | Horse Power...... | 1,57 | 3.68 | 6.24 | 11.05 | 19.63 | 44.21 | 78.53 | 122.87 | 176.86 |
| | Cubic Feet........ | 5.16 | 12.04 | 20 41 | 86.14 | 64.20 | 144.59 | 256.82 | 401.81 | 578.38 |
| 82 Lbs. | Miner's Inches... | 8.22 | 7.52 | 12.75 | 24.09 | 42.80 | 96.39 | 171.21 | 267.87 | 885.58 |
| 6632.86 | Revolutions........ | 2106 | 1053 | 843 | 702 | 527 | 851 | 263 | 211 | 176 |
| **200** | Horse Power...... | 1.70 | 3.97 | 6.74 | 11.93 | 21.20 | 47.75 | 84.81 | 132.70 | 191.00 |
| | Cubic Feet........ | 5.29 | 12.36 | 20.94 | 37.08 | 65.87 | 148.35 | 263.49 | 412.25 | 598.40 |
| 87 Lbs. | Miner's Inches... | 3.30 | 7.72 | 13.08 | 24.72 | 43.91 | 98.90 | 175.66 | 274.83 | 395.60 |
| 6805.17 | Revolutions........ | 2160 | 1080 | 864 | 720 | 540 | 360 | 270 | 216 | 180 |
| **210** | Horse Power...... | 1.83 | 4.28 | 7.25 | 12.84 | 22.81 | 51,38 | 91.26 | 142.78 | 205.52 |
| | Cubic Feet........ | 5.42 | 12.66 | 21.46 | 38.00 | 67.50 | 152.01 | 270.00 | 422.44 | 608.06 |
| 91 Lbs. | Miner's Inches... | 8.38 | 7.91 | 13.41 | 25.33 | 45.00 | 101.34 | 180.00 | 281.62 | 405.87 |
| 6973.26 | Revolutions....... | 2214 | 1107 | 885 | 738 | 553 | 369 | 277 | 222 | 184 |
| **220** | Horse Power..... | 1 96 | 4.59 | 7.77 | 13.77 | 24.46 | 55.09 | 97.85 | 153.10 | 220.86 |
| | Cubic Feet........ | 5.55 | 12.96 | 21.96 | 38.89 | 69.08 | 155.59 | 276.35 | 432.38 | 622.36 |
| 95 Lbs. | Miner's Inches.. | 3.46 | 8 10 | 13.72 | 25 93 | 46.05 | 103.72 | 184.23 | 288.25 | 414.91 |
| 7137.35 | Revolutions....... | 2268 | 1134 | 906 | 756 | 567 | 378 | 283 | 226 | 189 |
| **230** | Horse Power...... | 2.10 | 4.90 | 8.31 | 14.72 | 26.15 | 58.89 | 104.60 | 163.66 | 235.56 |
| | Cubic Feet........ | 5.68 | 13.25 | 22.46 | 39.77 | 70.64 | 159.08 | 282.56 | 442.09 | 636.35 |
| 100 Lbs. | Miner's Inches... | 3.55 | 8.28 | 14.03 | 26.51 | 47.09 | 106.06 | 188.38 | 294.73 | 424.24 |
| 7297.78 | Revolutions..... . | 2319 | 1159 | 928 | 778 | 580 | 386 | 290 | 232 | 193 |
| **240** | Horse Power..... | 2.24 | 5.23 | 8.86 | 15.69 | 27.87 | 62.77 | 111.50 | 174.45 | 251.10 |
| | Cubic Feet........ | 5.80 | 13.54 | 22.93 | 40.62 | 72.16 | 162.50 | 288.64 | 451.60 | 650.03 |
| 105 Lbs. | Miner's Inches... | 3.62 | 8.46 | 14.33 | 27.08 | 48.10 | 108.34 | 192.42 | 301.07 | 483.36 |
| 7454.70 | Revolutions........ | 2370 | 1185 | 948 | 790 | 592 | 395 | 296 | 287 | 197 |
| **250** | Horse Power...... | 2.38 | 5.56 | 9.42 | 16.68 | 29.63 | 66.74 | 118.54 | 185.47 | 266.96 |
| | Cubic Feet........ | 5.92 | 13.82 | 23.42 | 41.46 | 73.64 | 165.86 | 294.59 | 460.91 | 663.45 |
| 108 Lbs. | Miner's Inches... | 3.70 | 8.63 | 14.63 | 27.64 | 49.09 | 110.57 | 196.39 | 307.27 | 442.30 |
| 7608.41 | Revolutions........ | 2418 | 1209 | 966 | 806 | 605 | 403 | 302 | 241 | 202 |
| **260** | Horse Power...... | 2.52 | 5.89 | 10.05 | 17.69 | 31.43 | 70.78 | 125.72 | 196.71 | 283.15 |
| | Cubic Feet........ | 6.04 | 14.09 | 23.88 | 42.28 | 75.10 | 169.14 | 300.43 | 470.04 | 676.59 |
| 113 Lbs. | Miner's Inches... | 3.77 | 8.80 | 14.92 | 28.19 | 50.07 | 112.76 | 200.28 | 813.36 | 451.05 |
| 7759.10 | Revolutions........ | 2466 | 1233 | 986 | 822 | 617 | 411 | 308 | 247 | 206 |
| **270** | Horse Power...... | 2.67 | 6.24 | 10.67 | 18.72 | 33.26 | 74.90 | 133.05 | 208.17 | 299.63 |
| | Cubic Feet........ | 2.15 | 14.36 | 24.34 | 43.09 | 76.53 | 172.36 | 306.15 | 479.00 | 689.46 |
| 118 Lbs. | Miner's Inches... | 3.84 | 8.97 | 15.21 | 28.72 | 51.02 | 114.91 | 204.10 | 319 33 | 459.64 |
| 7906.93 | Revolutions........ | 2514 | 1257 | 1006 | 838 | 628 | 419 | 314 | 251 | 209 |
| **280** | Horse Power...... | 2.82 | 6.59 | 11.16 | 19.77 | 35.12 | 79.11 | 140.51 | 219.84 | 316.44 |
| | Cubic Feet........ | 6.26 | 14.62 | 24.79 | 43.88 | 77.94 | 175.53 | 311.77 | 487.79 | 702.12 |
| 121 Lbs. | Miner's Inches... | 3.91 | 9.13 | 15.49 | 29.25 | 51.29 | 117.02 | 205.18 | 825.19 | 468.06 |
| 8052.01 | Revolutions........ | 2562 | 1281 | 1025 | 854 | 639 | 427 | 319 | 255 | 213 |
| **290** | Horse Power...... | 2.97 | 6.94 | 11.77 | 20.84 | 37.02 | 83.38 | 148.10 | 231.73 | 333.55 |
| | Cubic Feet........ | 6.38 | 14.88 | 25.23 | 44.66 | 79.32 | 178.64 | 317.29 | 496.42 | 714.56 |
| 126 Lbs. | Miner's Inches... | 3.98 | 9.30 | 15.76 | 29.77 | 52.88 | 119.09 | 211.52 | 330.94 | 476.36 |
| 8194.54 | Revolutions........ | 2607 | 1303 | 1042 | 869 | 651 | 434 | 325 | 260 | 217 |
| **300** | Horse Power...... | 3.13 | 7.31 | 12.38 | 21.93 | 38.95 | 87.73 | 155.83 | 243.82 | 350.94 |
| | Cubic Feet........ | 6.48 | 15.13 | 25.66 | 45.42 | 80.67 | 181.69 | 322.71 | 504.91 | 726.76 |
| 130 Lbs. | Miner's Inches... | 4.05 | 9.45 | 16.03 | 30.28 | 53.78 | 121.12 | 215.14 | 336.60 | 484.51 |
| 8334.62 | Revolutions........ | 2652 | 1326 | 1060 | 884 | 663 | 442 | 331 | 265 | 221 |
| **310** | Horse Power...... | 3.29 | 7.68 | 13.01 | 23.04 | 40.92 | 92.16 | 163.69 | 256.11 | 368.64 |
| | Cubic Feet........ | 6.59 | 15.39 | 26.08 | 46.17 | 82.01 | 184.69 | 328.04 | 513.25 | 738.78 |
| 134 Lbs. | Miner's Inches... | 4.11 | 9.61 | 16.30 | 30.78 | 54.67 | 123.12 | 218.69 | 342.17 | 492.51 |
| 8472.39 | Revolutions........ | 2697 | 1348 | 1078 | 899 | 674 | 449 | 337 | 269 | 224 |

# PELTON WATER WHEEL TABLES.

**The calculations for power in these tables are based upon the application of one stream to the Wheel, as also upon an 85 per cent efficiency and effective heads, no allowance being made for loss of head by friction in pipe. The smaller figures under those denoting the various heads give the equivalent pressure in pounds and spouting velocity of the water in feet per minute. The cubic feet measurement is also based on the flow per minute.**

| HEAD IN FEET | SIZE OF WHEELS. | 6 Inch | 12 Inch | 15 Inch | 18 Inch | 24 Inch | 3 Foot | 4 Foot | 5 Foot | 6 Foot |
|---|---|---|---|---|---|---|---|---|---|---|
| **320** | Horse Power...... | 3.45 | 8.05 | 13.64 | 24.16 | 42.91 | 96.65 | 171.68 | 268.60 | 386.62 |
| 139 Lbs. | Cubic Feet......... | 6.70 | 15.63 | 26.50 | 46.91 | 83.32 | 187.65 | 333.29 | 521.46 | 750.60 |
| | Miner's Inches.. | 4.18 | 9.76 | 16.56 | 31.27 | 55.55 | 125.10 | 222.19 | 347.64 | 500.40 |
| 8607.94 | Revolutions........ | 2739 | 1369 | 1095 | 913 | 685 | 456 | 342 | 274 | 228 |
| **330** | Horse Power...... | 3.61 | 8.43 | 14.29 | 25.30 | 44.94 | 101.22 | 179.72 | 281.29 | 404.89 |
| 143 Lbs. | Cubic Feet......... | 6.80 | 15.88 | 26.91 | 47.64 | 84.61 | 190.56 | 338.46 | 529.55 | 762.24 |
| | Miner's Inches... | 4.25 | 9.92 | 16.81 | 31.76 | 56.41 | 127.04 | 225.64 | 353.03 | 508.16 |
| 8741.43 | Revolutions........ | 2781 | 1390 | 1111 | 927 | 696 | 463 | 346 | 277 | 231 |
| **340** | Horse Power...... | 3.78 | 8.82 | 14.94 | 26.46 | 47.00 | 105.86 | 188.02 | 294.18 | 423.44 |
| 147 Lbs. | Cubic Feet......... | 6.90 | 16.12 | 27.31 | 48.35 | 85.88 | 193.42 | 343.55 | 537.51 | 773.71 |
| | Miner's Inches... | 4.31 | 10.07 | 17.06 | 32.24 | 57.26 | 128.98 | 229.04 | 358.34 | 515.93 |
| 8872.89 | Revolutions........ | 2823 | 1411 | 1130 | 941 | 706 | 470 | 353 | 282 | 235 |
| **350** | Horse Power...... | 3.94 | 9.21 | 15.61 | 27.64 | 49.09 | 110.56 | 196.38 | 307.25 | 442.27 |
| 152 Lbs. | Cubic Feet......... | 7.00 | 16.35 | 27.71 | 49.06 | 87.14 | 196.25 | 348.57 | 545.36 | 785.00 |
| | Miner's Inches... | 4.37 | 10.21 | 17.32 | 32.70 | 58.09 | 130.83 | 232.38 | 363.57 | 523.32 |
| 9002.43 | Revolutions........ | 2865 | 1432 | 1146 | 955 | 716 | 477 | 358 | 285 | 238 |
| **360** | Horse Power...... | 4.10 | 9.61 | 16.28 | 28.83 | 51.21 | 115.34 | 204.86 | 320.52 | 461.36 |
| 156 Lbs. | Cubic Feet......... | 7.10 | 16.58 | 28.10 | 49.75 | 88.37 | 199.03 | 353.51 | 553.10 | 796.14 |
| | Miner's Inches... | 4.43 | 10.36 | 17.56 | 33.17 | 58.91 | 132.68 | 235.64 | 368.73 | 530.75 |
| 9130.14 | Revolutions........ | 2907 | 1453 | 1161 | 969 | 726 | 484 | 363 | 290 | 242 |
| **370** | Horse Power...... | 4.29 | 10.01 | 16.97 | 30.04 | 53.36 | 120.18 | 213.45 | 333.97 | 480.72 |
| 160 Lbs. | Cubic Feet......... | 7.20 | 16.81 | 28.49 | 50.44 | 89.59 | 201.78 | 358.39 | 560.73 | 807.12 |
| | Miner's Inches... | 4.50 | 10.50 | 17.75 | 33.63 | 59.73 | 134.52 | 238.92 | 373.82 | 538.08 |
| 9256.02 | Revolutions....... | 2946 | 1473 | 1178 | 982 | 736 | 491 | 368 | 294 | 245 |
| **380** | Horse Power...... | 4.46 | 10.42 | 17.66 | 31.27 | 55.54 | 125.08 | 222.16 | 347.60 | 500.33 |
| 165 Lbs. | Cubic Feet......... | 7.30 | 17.04 | 28.88 | 51.12 | 90.80 | 204.48 | 363.20 | 568.25 | 817.95 |
| | Miner's Inches... | 4.56 | 10.65 | 18.03 | 34.08 | 60.53 | 136.32 | 242.13 | 378.83 | 545.29 |
| 9380.32 | Revolutions..... .. | 2985 | 1492 | 1194 | 995 | 746 | 497 | 373 | 298 | 248 |
| **390** | Horse Power...... | 4.64 | 10.83 | 18.36 | 32.51 | 57.75 | 130.05 | 231.00 | 361.41 | 520.20 |
| 169 Lbs. | Cubic Feet......... | 7.39 | 17.26 | 29.25 | 51.79 | 91.98 | 207.16 | 367.95 | 575.68 | 828.64 |
| | Miner's Inches... | 4.61 | 10.78 | 18.25 | 34.53 | 61.32 | 138.11 | 245.28 | 383.79 | 552.43 |
| 9502.93 | Revolutions........ | 3024 | 1512 | 1210 | 1008 | 756 | 504 | 378 | 302 | 252 |
| **400** | Horse Power...... | 4.82 | 11.25 | 19.07 | 33.77 | 59.98 | 135.08 | 239.94 | 375.40 | 540.35 |
| 173 Lbs. | Cubic Feet......... | 7.49 | 17.48 | 29.63 | 52.45 | 93.16 | 209.80 | 372.64 | 583.02 | 839.20 |
| | Miner's Inches... | 4.68 | 10.92 | 18.51 | 34.96 | 62.10 | 139.84 | 248.40 | 388.68 | 559.35 |
| 9624.00 | Revolutions........ | 3063 | 1531 | 1225 | 1021 | 765 | 510 | 382 | 306 | 255 |
| **410** | Horse Power..... | 5.00 | 11.68 | 19.79 | 35.04 | 62.24 | 140.18 | 248.99 | 389.57 | 560.75 |
| 178 Lbs. | Cubic Feet........ | 7.58 | 17.70 | 30.00 | 53.10 | 94.31 | 212.40 | 377.26 | 590.26 | 849.63 |
| | Miner's Inches... | 4.73 | 11.06 | 18.75 | 35.40 | 62.87 | 141.60 | 251.51 | 393.50 | 566.41 |
| 9743.57 | Revolutions....... | 3102 | 1551 | 1240 | 1034 | 775 | 517 | 387 | 309 | 258 |
| **420** | Horse Power...... | 5.19 | 12.11 | 20.52 | 36.33 | 64.54 | 145.34 | 258.16 | 403.91 | 581.39 |
| 182 Lbs. | Cubic Feet......... | 7.67 | 17.91 | 30.36 | 53.74 | 95.46 | 214.98 | 381.84 | 597.41 | 859.93 |
| | Miner's Inches... | 4.79 | 11.19 | 18.93 | 35.83 | 63.64 | 143.32 | 254.56 | 398.28 | 573.28 |
| 9861.66 | Revolutions.. ..... | 3141 | 1570 | 1255 | 1047 | 785 | 523 | 392 | 313 | 261 |
| **430** | Horse Power...... | 5.37 | 12.54 | 21.26 | 37.64 | 66.86 | 150.57 | 267.44 | 418.42 | 602.28 |
| 186 Lbs. | Cubic Feet......... | 7.76 | 18.12 | 30.72 | 54.38 | 96.59 | 217.52 | 386.36 | 604.48 | 870.11 |
| | Miner's Inches... | 4.85 | 11.31 | 19.18 | 36.25 | 64.39 | 145.00 | 257.58 | 402.99 | 580.00 |
| 9978.35 | Revolutions........ | 3177 | 1588 | 1270 | 1059 | 794 | 529 | 397 | 317 | 264 |
| **440** | Horse Power...... | 5.56 | 12.98 | 22.01 | 38.96 | 69.20 | 155.85 | 276.82 | 433.11 | 623.40 |
| 191 Lbs. | Cubic Feet......... | 7.85 | 18.33 | 31.07 | 55.01 | 97.70 | 220.04 | 390.82 | 611.47 | 880.16 |
| | Miner's Inches... | 4.90 | 11.45 | 19.41 | 36.66 | 65.13 | 146.64 | 260.58 | 407.65 | 586.56 |
| 10093.74 | Revolutions........ | 3213 | 1606 | 1285 | 1071 | 803 | 535 | 401 | 320 | 267 |
| **450** | Horse Power...... | 5.75 | 13.43 | 22.76 | 40.29 | 71.57 | 161.19 | 286.31 | 447.95 | 644.78 |
| 195 Lbs. | Cubic Feet......... | 7.94 | 18.54 | 31.42 | 55.63 | 98.81 | 222.52 | 395.24 | 618.38 | 890.11 |
| | Miner's Inches... | 4.96 | 11.58 | 19.65 | 37.08 | 65.87 | 148.35 | 263.49 | 412.25 | 593 40 |
| 10207.79 | Revolutions........ | 3249 | 1624 | 1300 | 1083 | 812 | 541 | 406 | 324 | 270 |
| **460** | Horse Power...... | 5.95 | 13.88 | 23.53 | 41.65 | 73.97 | 166.60 | 295.91 | 462.97 | 666.40 |
| 200 Lbs. | Cubic Feet......... | 8.03 | 18.74 | 31.77 | 56.24 | 99.90 | 224.98 | 399.61 | 625.22 | 899.95 |
| | Miner's Inches... | 5.01 | 11.71 | 19.79 | 37.50 | 66.60 | 150.00 | 266.40 | 416.80 | 600.00 |
| 10320.58 | Revolutions........ | 3285 | 1642 | 1315 | 1095 | 821 | 547 | 410 | 327 | 273 |

# PELTON WATER WHEEL TABLES.

The calculations for power in these tables are based upon the application of one stream to the Wheel, as also upon an 85 per cent efficiency and effective heads, no allowance being made for loss of head by friction in pipe. The smaller figures under those denoting the various heads give the equivalent pressure in pounds and spouting velocity of the water in feet per minute. The cubic feet measurement is also based on the flow per minute.

| HEAD IN FEET | SIZE OF WHEELS. | 6 Inch | 12 Inch | 15 Inch | 18 Inch | 24 Inch | 3 Foot | 4 Foot | 5 Foot | 6 Foot |
|---|---|---|---|---|---|---|---|---|---|---|
| **470** | Horse Power...... | 6.14 | 14.33 | 24.29 | 43.01 | 76.40 | 172.06 | 305.61 | 478.15 | 688.25 |
| | Cubic Feet......... | 8.12 | 18.95 | 32.11 | 56.85 | 100.98 | 227.42 | 408.93 | 631.98 | 909.68 |
| 204 Lbs. | Miner's Inches.. | 5.07 | 11.84 | 20.06 | 37.90 | 67.32 | 151.61 | 269.28 | 421.32 | 606.44 |
| 10432.17 | Revolutions......... | 3321 | 1660 | 1328 | 1107 | 830 | 553 | 415 | 332 | 276 |
| **480** | Horse Power...... | 6.34 | 14.79 | 25.07 | 44.39 | 78.85 | 177.58 | 315.42 | 493.49 | 710.33 |
| | Cubic Feet......... | 8.20 | 19.15 | 32.45 | 57.45 | 102.05 | 229.82 | 408.20 | 638.66 | 919.29 |
| 208 Lbs. | Miner's Inches... | 5.12 | 11.96 | 20.28 | 38.30 | 68.00 | 153.20 | 272.12 | 425.78 | 612.80 |
| 10542.56 | Revolutions......... | 3357 | 1678 | 1343 | 1119 | 889 | 559 | 419 | 335 | 279 |
| **490** | Horse Power.... . | 6.54 | 15.26 | 25.80 | 45.79 | 81.33 | 183.16 | 325.32 | 509.00 | 782.65 |
| | Cubic Feet........ | 8.29 | 19.35 | 32.98 | 58.05 | 103.10 | 232.20 | 412.43 | 645.28 | 928.83 |
| 212 Lbs. | Miner's Inches... | 5.18 | 12.09 | 20.61 | 88.70 | 68.74 | 154.80 | 274.96 | 430.19 | 619.20 |
| 10651.79 | Revolutions.. .... | 3390 | 1695 | 1355 | 1130 | 848 | 565 | 424 | 339 | 282 |
| **500** | Horse Power...... | 6.74 | 15.73 | 20.66 | 47.20 | 83.83 | 188.80 | 335.34 | 524.66 | 755.20 |
| | Cubic Feet......... | 8.37 | 19.54 | 33.12 | 58.64 | 104.15 | 234.56 | 416.62 | 651.83 | 938 25 |
| 217 Lbs. | Miner's Inches... | 5.23 | 12.21 | 20.72 | 39.09 | 69.41 | 156.86 | 277.64 | 434.56 | 625.44 |
| 10759.96 | Revolutions.. ... | 3426 | 1713 | 1370 | 1142 | 856 | 571 | 428 | 342 | 285 |
| **520** | Horse Power...... | 7.06 | 19.57 | 28.16 | 50.05 | 88.90 | 200.22 | 355.62 | 556.39 | 800.88 |
| | Cubic Feet......... | 8.43 | 23.38 | 33.64 | 59.80 | 106.21 | 239.21 | 424.87 | 664.74 | 956.84 |
| 226 Lbs. | Miner's Inches... | 5.62 | 15.58 | 22.42 | 39.86 | 70.81 | 159.47 | 283.24 | 443.16 | 637.88 |
| 10973.04 | Revolutions......... | 3492 | 1746 | 1397 | 1164 | 873 | 582 | 436 | 349 | 291 |
| **540** | Horse Power...... | 7.47 | 20.72 | 29.80 | 52.97 | 94.08 | 211.88 | 376.33 | 588.80 | 847.52 |
| | Cubic Feet......... | 8.59 | 23.83 | 34.28 | 60.94 | 108.24 | 243.76 | 432.96 | 677.41 | 975.07 |
| 234 Lbs. | Miner's Inches... | 5.72 | 15.84 | 22.85 | 40 63 | 72.16 | 162.51 | 288.64 | 451.61 | 650.04 |
| 11182.07 | Revolutions. | 3561 | 1780 | 1424 | 1187 | 890 | 593 | 445 | 356 | 296 |
| **560** | Horse Power...... | 7.89 | 21.87 | 31.47 | 55.94 | 99.35 | 223.76 | 397.43 | 621.82 | 895.04 |
| | Cubic Feet......... | 8.75 | 24.26 | 34.91 | 62.06 | 110.23 | 248.24 | 440.91 | 689.84 | 992.96 |
| 243 Lbs. | Miner's Inches... | 5.72 | 16.17 | 23.27 | 41.37 | 73.48 | 165.49 | 293.94 | 459.89 | 661.96 |
| 11387.26 | Revolutions..... .. | 3624 | 1812 | 1450 | 1208 | 906 | 604 | 453 | 362 | 302 |
| **580** | Horse Power...... | 8.31 | 23.05 | 33.18 | 58.96 | 104.73 | 235.86 | 418.92 | 655.43 | 943.44 |
| | Cubic Feet......... | 8.90 | 24.69 | 35.53 | 63.16 | 112.18 | 252.63 | 448.71 | 702.04 | 1010.54 |
| 252 Lbs. | Miner's Inches... | 5.93 | 16.46 | 23.68 | 42.10 | 74 78 | 168.42 | 299.14 | 468.03 | 673.69 |
| 11588.83 | Revolutions....... | 3690 | 1844 | 1476 | 1230 | 922 | 615 | 461 | 369 | 307 |
| **600** | Horse Power..... | 8.75 | 24.26 | 34.90 | 62.04 | 110.19 | 248.16 | 440.77 | 689.63 | 992.65 |
| | Cubic Feet......... | 9.06 | 25.12 | 36.13 | 64.24 | 114.09 | 256.95 | 456.38 | 714.05 | 1027.80 |
| 260 Lbs. | Miner's Inches... | 6.04 | 16.74 | 24.08 | 42.82 | 76.06 | 171.30 | 304.24 | 476.03 | 685.20 |
| 11786.94 | Revolutions ....... | 3753 | 1876 | 1500 | 1251 | 938 | 625 | 469 | 375 | 312 |
| **650** | Horse Power..... | 9.87 | 27.35 | 39.36 | 69.95 | 124.25 | 279.82 | 497.01 | 777.62 | 1119.29 |
| | Cubic Feet......... | 9.43 | 26.14 | 37.61 | 66.86 | 118.75 | 267.44 | 475.02 | 743.21 | 1069.77 |
| 282 Lbs. | Miner's Inches... | 6.28 | 17.42 | 25.07 | 44.57 | 79.17 | 178.29 | 316.68 | 495.47 | 713.18 |
| 12268.24 | Revolutions.. .... | 3906 | 1952 | 1562 | 1302 | 976 | 651 | 488 | 390 | 325 |
| **700** | Horse Power...... | 11.02 | 30.57 | 44.00 | 78.18 | 138.86 | 312.73 | 555.46 | 869.06 | 1250.92 |
| | Cubic Feet......... | 9.78 | 27.13 | 39.03 | 69.38 | 123.23 | 277.54 | 492.95 | 771.26 | 1110.16 |
| 304 Lbs. | Miner's Inches... | 6.52 | 18.08 | 26.02 | 46.25 | 82.16 | 185.02 | 328.63 | 514.18 | 740.09 |
| 12731.34 | Revolutions.. .... | 4053 | 2026 | 1621 | 1351 | 1013 | 675 | 506 | 405 | 337 |
| **750** | Horse Power...... | 12.23 | 33.91 | 48.78 | 86.70 | 154.00 | 346.83 | 616.03 | 963.82 | 1387.34 |
| | Cubic Feet......... | 10.12 | 28.08 | 40.40 | 71.82 | 127.56 | 287.28 | 510.25 | 798.33 | 1149.13 |
| 326 Lbs. | Miner's Inches... | 6.75 | 18.72 | 26.93 | 47.88 | 85.04 | 191.52 | 340.16 | 532.22 | 766.09 |
| 13178.19 | Revolutions........ | 4197 | 2098 | 1680 | 1399 | 1049 | 699 | 524 | 419 | 349 |
| **800** | Horse Power...... | 13.47 | 37.35 | 53.73 | 95.52 | 169.66 | 382.09 | 678.66 | 1061.81 | 1528.36 |
| | Cubic Feet... .... | 10.46 | 29.00 | 41.72 | 74.17 | 131.74 | 296.70 | 526.99 | 824.51 | 1186.81 |
| 348 Lbs. | Miner's Inches... | 6.97 | 19.33 | 27.81 | 49.45 | 87.83 | 197.80 | 351.82 | 549.68 | 791.21 |
| 13610.40 | Revolutions.. .... | 4332 | 2166 | 1732 | 1444 | 1083 | 722 | 542 | 433 | 361 |
| **900** | Horse Power...... | 16.07 | 44.57 | 64.12 | 113.98 | 202.45 | 455.94 | 809.82 | 1267.02 | 1823.76 |
| | Cubic Feet......... | 11.09 | 30.76 | 44.25 | 78.67 | 139.74 | 314.70 | 558.96 | 874.53 | 1258 81 |
| 391 Lbs. | Miner's Inches... | 7.39 | 20.50 | 29.50 | 52.45 | 93.16 | 209.80 | 372.64 | 583.02 | 839.20 |
| 14436.00 | Revolutions......... | 4596 | 2298 | 1838 | 1532 | 1149 | 766 | 574 | 459 | 383 |
| **1000** | Horse Power...... | 18.82 | 52.20 | 75.11 | 133.50 | 237.12 | 534.01 | 948.48 | 1483.97 | 2136.04 |
| | Cubic Feet......... | 11.69 | 32.42 | 46.65 | 82.93 | 147.30 | 331.72 | 589.19 | 921.83 | 1326.91 |
| 434 Lbs. | Miner's Inches... | 7.79 | 21.61 | 31.10 | 55.29 | 98.20 | 221.15 | 392.79 | 614.56 | 884.61 |
| 15216.89 | Revolutions. ...... | 4845 | 2420 | 1938 | 1615 | 1210 | 807 | 605 | 484 | 403 |

## PELTON QUINTEX NOZZLE WHEEL.

The above cut illustrates what is known as the Quintex Nozzle Wheel, which has been desinged to meet the demand for a wheel to handle large quantities of water under moderately low heads. This style of nozzle has five rectangular openings, all controlled by a slide gate, by means of which any or all of the openings may be closed, thus affording a wide range of capacity and adaptation to a varying water supply, without any loss in efficiency. These wheels, as will be seen from tables, can be operated under any head to which a turbine is applicable, while they are much more efficient and reliable in every way, affording the same advantages in these respects as the STANDARD PELTON WHEEL.

This type of wheel has largely displaced all forms of turbines in Mexico, Central and South America, for running coffee and sugar machinery, as well as for various other purposes. They can be depended upon for the highest useful effect under all conditions of service, and in the most simple and economical way.

The Quintex Nozzle Wheel is intended to be mounted on wood frame or concrete foundations as may be desired. It is also made with cast-iron bedplate and iron housings, thus being entirely self-contained; a form of construction intended more particularly for localities where woodwork is difficult to obtain, or climatic conditions unfavorable to its use. On the following page will be found tables giving the capacity, speed, and prices of the various sizes of these wheels under given heads. The prices therein are based on wheels for mounting on wood frames or masonry foundation. The framework is not included, but furnished when desired, with wheel properly fitted, then all taken down and packed for shipment. For full iron-cased wheels as above described, 50 per cent should be added to prices given.

## PRICE LIST, POWER AND WEIGHT QUINTEX NOZZLE WHEELS

| HEAD IN FEET | SIZE OF WHEELS | 24 in. | 42 in. | 48 in. | 24 in. Prices | 42 in. Prices | 48 in. Prices |
|---|---|---|---|---|---|---|---|
| | Horse Power | 1.61 | 6.45 | 10 | | | |
| 10 | Cubic Feet | 107 | 428 | 668 | $225 | $450 | $750 |
| | Revolutions | 121 | 69 | 61 | | | |
| | Horse Power | 2.97 | 11.94 | 18.55 | | | |
| 15 | Cubi Feet | 131 | 526 | 817 | 225 | 450 | 750 |
| | Revolutions | 148 | 85 | 74 | | | |
| | Horse P wer | 3.17 | 18.27 | 28.5 | | | |
| 20 | Cubic Feet | 151 | 605 | 944 | 225 | 450 | 750 |
| | Revolutions | 171 | 98 | 85 | | | |
| | Horse Pow r | 6.37 | 25.5 | 39.7 | | | |
| 25 | Cubic Feet | 169 | 677 | 1052 | 225 | 480 | 750 |
| | Revolutions | 191 | 109 | 96 | | | |
| | Horse Power | 8.38 | 33.6 | 52.3 | | | |
| 30 | Cubic Feet | 185 | 742 | 1154 | 225 | 480 | 775 |
| | Revolutions | 210 | 119 | 105 | | | |
| | Horse Power | 10.5 | 42.2 | 65.7 | | | |
| 35 | Cubic Feet | 200 | 800 | 1246 | 225 | 480 | 775 |
| | Revolutions | 226 | 129 | 113 | | | |
| | Horse Power | 12.9 | 51.7 | 80.4 | | | |
| 40 | Cubic Feet | 214 | 857 | 1331 | 225 | 480 | 775 |
| | Revolutions | 242 | 138 | 121 | | | |
| | Horse Power | 15.4 | 61.7 | 96 | | | |
| 45 | Cubic Feet | 227 | 908 | 1412 | 225 | 500 | 775 |
| | Revolutions | 257 | 147 | 129 | | | |
| | Horse Power | 18 | 72.4 | 112.5 | | | |
| 50 | Cubic Feet | 239 | 958 | 1492 | 225 | 500 | 775 |
| | Revolutions | 271 | 155 | 135 | | | |
| | Horse Power | 20.8 | 83.5 | 129.6 | | | |
| 55 | Cubic Feet | 251 | 1005 | 1562 | 240 | 500 | 825 |
| | Revolutions | 284 | 162 | 142 | | | |
| | Horse Power | 23.7 | 94.9 | 147.8 | | | |
| 60 | Cubic Feet | 262 | 1048 | 1632 | 240 | 500 | 825 |
| | Revolutions | 296 | 169 | 148 | | | |
| | Horse Power | 26.7 | 107 | 166.5 | | | |
| 65 | Cubic Feet | 273 | 1092 | 1698 | 240 | 525 | 825 |
| | Revolutions | 309 | 176 | 154 | | | |
| | Horse Power | 29.9 | 119.6 | 186.5 | | | |
| 70 | Cubic Feet | 283 | 1132 | 1764 | 240 | 525 | 875 |
| | Revolutions | 320 | 183 | 160 | | | |
| | Horse Power | 33.2 | 132.7 | 206.5 | | | |
| 75 | Cubic Feet | 293 | 1172 | 1825 | 240 | 525 | 875 |
| | Revolutions | 332 | 189 | 166 | | | |
| | Horse Power | 36.6 | 146.4 | 228 | | | |
| 80 | Cubic Feet | 303 | 1213 | 1886 | 250 | 525 | 875 |
| | Revolutions | 342 | 196 | 171 | | | |
| | Horse Power | 40 | 160 | 249 | | | |
| 85 | Cubic Feet | 312 | 1248 | 1943 | 250 | 550 | 975 |
| | Revolutions | 353 | 202 | 177 | | | |
| | Horse Power | 43.5 | 174.8 | 271.8 | | | |
| 90 | Cubic Feet | 321 | 1285 | 1997 | 250 | 550 | 975 |
| | Revolutions | 363 | 208 | 182 | | | |
| | Horse Power | 47.4 | 189.5 | 291.7 | | | |
| 95 | Cubic Feet | 330 | 1320 | 2054 | 250 | 550 | 975 |
| | Revolutions | 373 | 213 | 186 | | | |
| | Horse Power | 51.2 | 204.4 | 318.1 | | | |
| 100 | Cubic Feet | 339 | 1354 | 2107 | 250 | 550 | 975 |
| | Revolutions | 383 | 219 | 192 | | | |
| | | | | | WEIGHT 800-1400 LBS. | WEIGHT 2500-3500 LBS. | WEIGHT 6500-7500 LBS. |

## TABLES FOR CALCULATING THE HORSE POWER OF WATER.

### MINERS' INCH TABLE.

The following table gives the Horse-Power of one miner's inch of water under heads from one up to eleven hundred feet. This inch equals 1½ cubic feet per minute.

### CUBIC FEET TABLE.

The following table gives the Horse-Power of one cubic foot of water per minute under heads from one up to eleven hundred feet.

| Heads in Feet. | Horse Power. | Heads in Feet. | Horse Power. | Heads in Feet. | Horse Power. | Heads in Feet. | Horse Power. |
|---|---|---|---|---|---|---|---|
| 1 | .0024147 | 320 | .772704 | 1 | .0016098 | 320 | .515136 |
| 20 | .0482294 | 330 | .796851 | 20 | .032196 | 330 | .531234 |
| 30 | .072441 | 340 | .820998 | 30 | .048294 | 340 | .547332 |
| 40 | .096588 | 350 | .845145 | 40 | .064392 | 350 | .563430 |
| 50 | .120735 | 360 | .869292 | 50 | .080490 | 360 | .579528 |
| 60 | .144882 | 370 | .893439 | 60 | .096588 | 370 | .595626 |
| 70 | .169029 | 380 | .917586 | 70 | .112686 | 380 | .611724 |
| 80 | .193176 | 390 | .941733 | 80 | .128784 | 390 | .627822 |
| 90 | .217323 | 400 | .965880 | 90 | .144892 | 400 | .643920 |
| 100 | .241470 | 410 | .990027 | 100 | .160980 | 410 | .660018 |
| 110 | .265617 | 420 | 1.014174 | 110 | .177078 | 420 | .676116 |
| 120 | .289764 | 430 | 1.038321 | 120 | .193176 | 430 | .692214 |
| 130 | .313911 | 440 | 1.062468 | 130 | .209274 | 440 | .708312 |
| 140 | .338058 | 450 | 1.086615 | 140 | .225372 | 450 | .724410 |
| 150 | .362205 | 460 | 1.110762 | 150 | .241470 | 460 | .740508 |
| 160 | .386352 | 470 | 1.134909 | 160 | .257568 | 470 | .756606 |
| 170 | .410499 | 480 | 1.159056 | 170 | .273666 | 480 | .772704 |
| 180 | .434646 | 490 | 1.183206 | 180 | .289764 | 490 | .788802 |
| 190 | .458793 | 500 | 1.207350 | 190 | .305862 | 500 | .804900 |
| 200 | .482940 | 520 | 1.255644 | 200 | .321960 | 520 | .837096 |
| 210 | .507087 | 540 | 1.303938 | 210 | .338058 | 540 | .869292 |
| 220 | .531234 | 560 | 1.352232 | 220 | .354156 | 560 | .901488 |
| 230 | .555381 | 580 | 1.400526 | 230 | .370254 | 580 | .933684 |
| 240 | .579528 | 600 | 1.448820 | 240 | .386352 | 600 | .965880 |
| 250 | .603675 | 650 | 1.569555 | 250 | .402450 | 650 | 1.046370 |
| 260 | .627822 | 700 | 1.690290 | 260 | .418548 | 700 | 1.126860 |
| 270 | .651969 | 750 | 1.811025 | 270 | .434646 | 750 | 1.207350 |
| 280 | .676116 | 800 | 1.931760 | 280 | .450744 | 800 | 1.287840 |
| 290 | .700263 | 900 | 2.173230 | 290 | .466842 | 900 | 1.448820 |
| 300 | .724410 | 1000 | 2.414700 | 300 | .482940 | 1000 | 1.609800 |
| 310 | .748557 | 1100 | 2.656170 | 310 | .499038 | 1100 | 1.770780 |

### WHEN THE EXACT HEAD IS FOUND IN ABOVE TABLE.

EXAMPLE.—Have 100 foot head and 50 inches of water. How many Horse-Power?

By reference to above table the Horse Power of 1 inch under 100 ft. head is .241470. This amount multiplied by the number of inches, 50, will give 12.07 Horse Power.

### WHEN EXACT HEAD IS NOT FOUND IN TABLE.

Take the Horse Power of 1 inch under 1 ft. head and multiply by the number of inches, and then by number of feet head. The product will be the required Horse Power.

The above formula will answer for the cubic feet table, by substituting the the equivalents therein for those of miner's inches.

NOTE.—The above tables are based upon an efficiency of 85%.

## TABLE OF RIVETED HYDRAULIC PIPE

**Showing Price and Weight, with safe head for various sizes of double riveted pipe.**

| Diameter of pipe in inches. | Area of pipe in inches. | Thickn's of iron by wire gauge. | Head in feet the pipe will safely stand. | Cub. ft. of water pipe will convey per min. at vel. 3 ft. per second. | Weight per lineal foot in lbs. | Price per Foot. | Diameter of pipe in inches. | Area of pipe in inches. | Thickn's of iron by wire gauge. | Head in feet the pipe will safely stand. | Cub. ft. of water pipe will convey per min. at vel. 3 ft. per second. | Weight per lineal foot in lbs. | Price per Foot. |
|---|---|---|---|---|---|---|---|---|---|---|---|---|---|
| 3 | 7 | 18 | 400 | 9 | 2 | $.20 | 18 | 254 | 16 | 165 | 320 | 16½ | $ 1.20 |
| 4 | 12 | 18 | 350 | 16 | 2¼ | .25 | 18 | 254 | 14 | 252 | 320 | 20½ | 1.40 |
| 4 | 12 | 16 | 525 | 16 | 3 | .35 | 18 | 254 | 12 | 385 | 320 | 27¼ | 1.90 |
| 5 | 20 | 18 | 325 | 25 | 3½ | .35 | 18 | 254 | 11 | 424 | 320 | 30 | 2.10 |
| 5 | 20 | 16 | 500 | 25 | 4½ | .45 | 18 | 254 | 10 | 505 | 320 | 34 | 2.40 |
| 5 | 20 | 14 | 675 | 25 | 5 | .50 | 20 | 314 | 16 | 148 | 400 | 18 | 1.26 |
| 6 | 28 | 18 | 296 | 36 | 4¼ | .44 | 20 | 314 | 14 | 227 | 400 | 22½ | 1.54 |
| 6 | 28 | 16 | 487 | 36 | 5¾ | .50 | 20 | 314 | 12 | 346 | 400 | 30 | 2.10 |
| 6 | 28 | 14 | 743 | 36 | 7½ | .56 | 20 | 314 | 11 | 380 | 400 | 32½ | 2.25 |
| 7 | 38 | 18 | 254 | 50 | 5¼ | .50 | 20 | 314 | 10 | 456 | 400 | 36½ | 2.50 |
| 7 | 38 | 16 | 419 | 50 | 6¾ | .56 | 22 | 380 | 16 | 135 | 480 | 20 | 1.40 |
| 7 | 38 | 14 | 640 | 50 | 8½ | .63 | 22 | 380 | 14 | 206 | 480 | 24¾ | 1.70 |
| 8 | 50 | 16 | 367 | 63 | 7½ | .65 | 22 | 380 | 12 | 316 | 480 | 32¾ | 2.25 |
| 8 | 50 | 14 | 560 | 63 | 9½ | .75 | 22 | 380 | 11 | 347 | 480 | 35¾ | 2.45 |
| 8 | 50 | 12 | 854 | 63 | 13 | .94 | 22 | 380 | 10 | 415 | 480 | 40 | 2.80 |
| 9 | 63 | 16 | 327 | 80 | 8½ | .69 | 24 | 452 | 14 | 188 | 570 | 27¼ | 1.80 |
| 9 | 63 | 14 | 499 | 80 | 10¾ | .88 | 24 | 452 | 12 | 290 | 570 | 35½ | 2.35 |
| 9 | 63 | 12 | 761 | 80 | 14¼ | 1.06 | 24 | 452 | 11 | 318 | 570 | 39 | 2.70 |
| 10 | 78 | 16 | 295 | 100 | 9¼ | .72 | 24 | 452 | 10 | 379 | 570 | 43½ | 2.95 |
| 10 | 78 | 14 | 450 | 100 | 11¾ | .82 | 24 | 452 | 8 | 466 | 570 | 53 | 3.50 |
| 10 | 78 | 12 | 637 | 100 | 15¾ | 1.00 | 26 | 530 | 14 | 175 | 670 | 29¼ | 2.00 |
| 10 | 78 | 11 | 754 | 100 | 17½ | 1.25 | 26 | 530 | 12 | 267 | 670 | 38½ | 2.[illegible] |
| 10 | 78 | 10 | 900 | 100 | 19¼ | 1.50 | 26 | 530 | 11 | 294 | 670 | 42 | 2.87 |
| 11 | 95 | 16 | 269 | 120 | 9¾ | .75 | 26 | 530 | 10 | 352 | 670 | 47 | 3.10 |
| 11 | 95 | 14 | 412 | 120 | 13 | .94 | 26 | 530 | 8 | 432 | 670 | 57¼ | 3.85 |
| 11 | 95 | 12 | 626 | 120 | 17¼ | 1.25 | 28 | 615 | 14 | 102 | 775 | 31½ | 2.12 |
| 11 | 95 | 11 | 687 | 120 | 18¾ | 1.44 | 28 | 615 | 12 | 247 | 775 | 41¼ | 2.75 |
| 11 | 95 | 10 | 820 | 120 | 21 | 1.62 | 28 | 615 | 11 | 273 | 775 | 45 | 3.00 |
| 12 | 113 | 16 | 246 | 142 | 11¼ | .82 | 28 | 615 | 10 | 327 | 775 | 50¼ | 3.20 |
| 12 | 113 | 14 | 377 | 142 | 14 | 1.00 | 28 | 615 | 8 | 400 | 775 | 61¼ | 4.15 |
| 12 | 113 | 12 | 574 | 142 | 18½ | 1.38 | 30 | 706 | 12 | 231 | 890 | 44 | 2.90 |
| 12 | 113 | 11 | 630 | 142 | 19¾ | 1.50 | 30 | 706 | 11 | 254 | 890 | 48 | 3.15 |
| 12 | 113 | 10 | 753 | 142 | 22¾ | 1.69 | 30 | 706 | 10 | 304 | 890 | 54 | 3.50 |
| 13 | 132 | 16 | 228 | 170 | 12 | .90 | 30 | 706 | 8 | 375 | 890 | 65 | 4.30 |
| 13 | 132 | 14 | 348 | 170 | 15 | 1.12 | 30 | 706 | 7 | 425 | 890 | 74 | 4.75 |
| 13 | 132 | 12 | 530 | 170 | 20 | 1.50 | 33 | 1017 | 11 | 141 | 1300 | 58 | 3.80 |
| 13 | 132 | 11 | 583 | 170 | 22 | 1.65 | 33 | 1017 | 10 | 155 | 1300 | 67 | 4.30 |
| 13 | 132 | 10 | 696 | 170 | 24½ | 1.80 | 36 | 1017 | 8 | 192 | 1300 | 78 | 5.10 |
| 14 | 153 | 16 | 211 | 200 | 13 | .98 | 36 | 1017 | 7 | 210 | 1300 | 88 | 5.75 |
| 14 | 153 | 14 | 324 | 200 | 16 | 1.17 | 40 | 1256 | 10 | 141 | 1600 | 71 | 4.75 |
| 14 | 153 | 12 | 494 | 200 | 21½ | 1.57 | 40 | 1256 | 8 | 174 | 1600 | 86 | 5.60 |
| 14 | 153 | 11 | 543 | 200 | 23½ | 1.72 | 40 | 1256 | 7 | 189 | 1600 | 97 | 6.40 |
| 14 | 153 | 10 | 648 | 200 | 26 | 1.95 | 40 | 1256 | 6 | 213 | 1600 | 108 | 7.35 |
| 15 | 176 | 16 | 197 | 225 | 13¾ | .96 | 40 | 1256 | 4 | 250 | 1600 | 126 | 8.50 |
| 15 | 176 | 14 | 302 | 225 | 17 | 1.28 | 42 | 1385 | 10 | 135 | 1760 | 74½ | 5.05 |
| 15 | 176 | 12 | 460 | 225 | 23 | 1.75 | 42 | 1385 | 8 | 165 | 1760 | 91 | 6.20 |
| 15 | 176 | 11 | 507 | 225 | 24½ | 1.95 | 42 | 1385 | 7 | 180 | 1760 | 102 | 7.00 |
| 15 | 176 | 10 | 606 | 225 | 28 | 2.10 | 42 | 1385 | 6 | 210 | 1760 | 114 | 7.80 |
| 16 | 201 | 16 | 185 | 255 | 14½ | 1.05 | 42 | 1385 | 4 | 240 | 1760 | 133 | 9.00 |
| 16 | 201 | 14 | 283 | 255 | 17¼ | 1.20 | 42 | 1385 | ¼ | 270 | 1760 | 137 | 9.50 |
| 16 | 201 | 12 | 432 | 255 | 24¼ | 1.70 | 42 | 1385 | 3 | 300 | 1760 | 145 | 10.00 |
| 16 | 201 | 11 | 474 | 255 | 26½ | 1.85 | 42 | 1385 | 5/16 | 321 | 1760 | 177 | 12.00 |
| 16 | 201 | 10 | 567 | 255 | 29½ | 2.00 | 42 | 1385 | 3/8 | 363 | 1760 | 216 | 15.00 |

NOTE.—Where formed and punched including rivets, for mule packing or to facilitate transportation by other means, 30 per cent. may be deducted from prices above given. This list is based upon pipe coated inside and out with asphaltum, and is given for the purpose of enabling parties to make an approximate estimate of the cost. Net prices will be quoted on application.

## LOSS OF HEAD IN PIPE BY FRICTION.

The following tables show the loss of head by friction in each 100 feet in length of different diameters of pipe when discharging the following quantities of water per minute:

INSIDE DIAMETER OF PIPE IN INCHES.

| | 1 | | 2 | | 3 | | 4 | | 5 | | 6 | |
|---|---|---|---|---|---|---|---|---|---|---|---|---|
| Velo in ft. per sec. | Loss of head in feet. | Cubic feet per min. | Loss of head in feet. | Cubic feet per min. | Loss of head in feet | Cubic feet per min. | Loss of head in feet. | Cubic feet per min. | Loss of head in feet. | Cubic feet per min. | Loss of head in feet. | Cubic feet per min. |
| 2.0 | 2.37 | .65 | 1.185 | 2.62 | .791 | 5.89 | .593 | 10.4 | .474 | 16.3 | .395 | 23.5 |
| 2.2 | 2.80 | .73 | 1.404 | 2.88 | .936 | 6.48 | .702 | 11.5 | .561 | 18. | .468 | 25.9 |
| 2.4 | 3.27 | .79 | 1.639 | 3.14 | 1.093 | 7.07 | .819 | 12.5 | .650 | 19.6 | .547 | 28.2 |
| 2.6 | 3.78 | .86 | 1.891 | 3.40 | 1.26 | 7.65 | .945 | 13.6 | .757 | 21.3 | .631 | 30.6 |
| 2.8 | 4.32 | .92 | 2.16 | 3.66 | 1.44 | 8.24 | 1.080 | 14.6 | .864 | 22.9 | .720 | 32.9 |
| 3.0 | 4.89 | .99 | 2.44 | 3.92 | 1.62 | 8.83 | 1.22 | 15.7 | .978 | 24.5 | .815 | 35.3 |
| 3.2 | 5.47 | 1.06 | 2.73 | 4.18 | 1.82 | 9.42 | 1.37 | 16.7 | 1.098 | 26.2 | .915 | 37.7 |
| 3.4 | 6.09 | 1.12 | 3.05 | 4.45 | 2.04 | 10.00 | 1.52 | 17.8 | 1.22 | 27.8 | 1.021 | 40. |
| 3.6 | 6.76 | 1.19 | 3.38 | 4.71 | 2.26 | 10.60 | 1.69 | 18.8 | 1.35 | 29.4 | 1.131 | 42.4 |
| 3.8 | 7.48 | 1.26 | 3.74 | 4.97 | 2.49 | 11.20 | 1.87 | 19.9 | 1.49 | 31. | 1.25 | 44.7 |
| 4.0 | 8.20 | 1.32 | 4.10 | 5.23 | 2.73 | 11.80 | 2.05 | 20.9 | 1.64 | 32.7 | 1.37 | 47.1 |
| 4.2 | 8.97 | 1.39 | 4.49 | 5.49 | 2.98 | 12.30 | 2.24 | 22.0 | 1.79 | 34.3 | 1.49 | 49.5 |
| 4.4 | 9.77 | 1.45 | 4.89 | 5.76 | 3.25 | 12.90 | 2.43 | 23.0 | 1.95 | 36.0 | 1.62 | 51.8 |
| 4.6 | 10.60 | 1.52 | 5.30 | 6.02 | 3.53 | 13.50 | 2.64 | 24.0 | 2.11 | 37.6 | 1.76 | 54.1 |
| 4.8 | 11.45 | 1.58 | 5.72 | 6.28 | 3.81 | 14.10 | 2.85 | 25.1 | 2.27 | 39.2 | 1.90 | 56.5 |
| 5.0 | 12.33 | 1.65 | 6.17 | 6.54 | 4.11 | 14.70 | 3.08 | 26.2 | 2.46 | 40.9 | 2.05 | 58.9 |
| 5.2 | 13.24 | 1.72 | 6.62 | 6.80 | 4.41 | 15.30 | 3.31 | 27.2 | 2.65 | 42.5 | 2.21 | 61.2 |
| 5.4 | 14.20 | 1.78 | 7.10 | 7.06 | 4.73 | 15.90 | 3.55 | 28.2 | 2.84 | 44.2 | 2.37 | 63.6 |
| 5.6 | 15.16 | 1.85 | 7.58 | 7.32 | 5.06 | 16.50 | 3.79 | 29.3 | 3.03 | 45.8 | 2.53 | 65.9 |
| 5.8 | 16.17 | 1.91 | 8.09 | 7.58 | 5.40 | 17.10 | 4.04 | 30.3 | 3.24 | 47.4 | 2.70 | 68.3 |
| 6.0 | 17.23 | 1.98 | 8.61 | 7.85 | 5.74 | 17.70 | 4.31 | 31.4 | 3.45 | 49.1 | 2.87 | 70.7 |
| 7.0 | 22.89 | 2.31 | 11.45 | 9.16 | 7.62 | 20.6 | 5.72 | 36.6 | 4.57 | 57.2 | 3.81 | 82.4 |

INSIDE DIAMETER OF PIPE IN INCHES.

| | 7 | | 8 | | 9 | | 10 | | 11 | | 12 | |
|---|---|---|---|---|---|---|---|---|---|---|---|---|
| Velo in ft. per sec. | Loss of head in feet. | Cubic feet per min. | Loss of head in feet. | Cubic feet per min. | Loss of head in feet. | Cubic feet per min. | Loss of head in feet. | Cubic feet per min. | Loss of head in feet. | Cubic feet per min. | Loss of head in feet. | Cubic feet per min. |
| 2.0 | .338 | 32.0 | .296 | 41.9 | .264 | 53. | .237 | 65.4 | .216 | 79.2 | .198 | 94.2 |
| 2.2 | .401 | 35.3 | .351 | 46.1 | .312 | 58.3 | .281 | 72. | .255 | 87.1 | .234 | 103. |
| 2.4 | .468 | 38.5 | .410 | 50.2 | .365 | 63.6 | .327 | 78.5 | .297 | 95.0 | .273 | 113. |
| 2.6 | .540 | 41.7 | .473 | 54.4 | .420 | 68.9 | .378 | 85.1 | .344 | 103. | .315 | 122. |
| 2.8 | .617 | 44.9 | .540 | 58.6 | .480 | 74.2 | .432 | 91.6 | .392 | 111. | .360 | 132. |
| 3.0 | .698 | 48.1 | .611 | 62.8 | .544 | 79.5 | .488 | 98.2 | .444 | 119. | .407 | 141. |
| 3.2 | .785 | 51.3 | .686 | 67. | .609 | 84.8 | .549 | 105. | .499 | 127. | .457 | 151. |
| 3.4 | .875 | 54.5 | .765 | 71.2 | .680 | 90.1 | .612 | 111. | .557 | 134. | .510 | 160. |
| 3.6 | .969 | 57.7 | .848 | 75.4 | .755 | 95.4 | .679 | 118. | .617 | 142. | .566 | 169. |
| 3.8 | 1.070 | 60.9 | .936 | 79.6 | .831 | 101. | .749 | 124. | .680 | 150. | .624 | 179. |
| 4.0 | 1.175 | 64.1 | 1.027 | 83.7 | .913 | 106. | .822 | 131. | .747 | 158. | .685 | 188. |
| 4.2 | 1.28 | 67.3 | 1.122 | 87.9 | .998 | 111. | .897 | 137. | .816 | 166. | .749 | 198. |
| 4.4 | 1.39 | 70.5 | 1.22 | 92.1 | 1.086 | 116. | .977 | 144. | .888 | 174. | .815 | 207. |
| 4.6 | 1.51 | 73.7 | 1.32 | 96.3 | 1.177 | 122. | 1.059 | 150. | .963 | 182. | .883 | 217. |
| 4.8 | 1.63 | 76.9 | 1.43 | 100.0 | 1.27 | 127. | 1.145 | 157. | 1.040 | 190. | .954 | 226 |
| 5.0 | 1.76 | 80.2 | 1.54 | 105. | 1.37 | 132. | 1.23 | 163. | 1.122 | 198. | 1.028 | 235. |
| 5.2 | 1.89 | 83.3 | 1.65 | 109. | 1.47 | 138. | 1.32 | 170. | 1.20 | 206. | 1.104 | 245 |
| 5.4 | 2.03 | 86.6 | 1.77 | 113. | 1.57 | 143. | 1.41 | 177. | 1.28 | 214. | 1.183 | 254. |
| 5.6 | 2.17 | 89.8 | 1.89 | 117. | 1.68 | 148. | 1.51 | 183. | 1.37 | 222. | 1.26 | 264. |
| 5.8 | 2.31 | 93.0 | 2.01 | 121. | 1.80 | 154. | 1.61 | 190. | 1.46 | 229. | 1.34 | 273. |
| 6.0 | 2.46 | 96.2 | 2.15 | 125. | 1.92 | 159. | 1.71 | 196. | 1.56 | 237. | 1.43 | 283. |
| 7.0 | 3.26 | 112.0 | 2.85 | 146. | 2.52 | 185. | 2.28 | 229. | 2.07 | 277. | 1.91 | 330. |

EXAMPLE.—Have 200 ft. head and 600 ft. of 11 inch pipe, carrying 119 cubic feet of water per minute. To find effective head. In right hand column under 11 inch pipe, find 119 cubic ft., opposite this will be found the co-efficient of friction for this amount of water, which is .444. Multiply this by the number of hundred feet of pipe, which is 6, and you will have 2.66 ft., which is the loss of head. Therefore the effective head is 200—2.66=197.34.

# LOSS OF HEAD IN PIPE BY FRICTION.

**The following tables show the loss of head by friction in each 100 feet in length of different diameters of pipe when discharging the following quantities of water per minute:**

INSIDE DIAMETER OF PIPE IN INCHES.

| | 13 | | 14 | | 15 | | 16 | | 18 | | 20 | |
|---|---|---|---|---|---|---|---|---|---|---|---|---|
| Velo in ft. per sec. | Loss of head in feet. | Cubic feet per min. | Loss of head in feet. | Cubic feet per min. | Loss of head in feet. | Cubic feet per min. | Loss of head in feet. | Cubic feet per min. | Loss of head in feet. | Cubic feet per min. | Loss of head in feet. | Cubic feet per min. |
| 2.0 | .183 | 110. | .169 | 128. | .158 | 147. | .147 | 167. | .132 | 212. | .119 | 262. |
| 2.2 | .216 | 121. | .200 | 141. | .187 | 162. | .175 | 184. | .156 | 233. | .140 | 288. |
| 2.4 | .252 | 133. | .234 | 154. | .218 | 176. | .205 | 201. | .182 | 254. | .164 | 314. |
| 2.6 | .290 | 144. | .270 | 167. | .252 | 191. | .236 | 218. | .210 | 275. | .189 | 340. |
| 2.8 | .332 | 156. | .308 | 179. | .288 | 206. | .270 | 234. | .240 | 297. | .216 | 366. |
| 3.0 | .375 | 166. | .349 | 192. | .325 | 221. | .306 | 251. | .271 | 318. | .245 | 393. |
| 3.2 | .422 | 177. | .392 | 205. | .366 | 235. | .343 | 268. | .305 | 339. | .275 | 419. |
| 3.4 | .471 | 188. | .438 | 218. | .408 | 250. | .383 | 284. | .339 | 360. | .306 | 445. |
| 3.6 | .522 | 199. | .485 | 231. | .452 | 265. | .425 | 301. | .377 | 382. | .339 | 471. |
| 3.8 | .576 | 210. | .535 | 243. | .499 | 280. | .468 | 318. | .416 | 403. | .374 | 497. |
| 4.0 | .632 | 221. | .587 | 256. | .548 | 294. | .513 | 335. | .456 | 424. | .410 | 523. |
| 4.2 | .691 | 232. | .641 | 269. | .598 | 309. | .561 | 352. | .499 | 445. | .449 | 550. |
| 4.4 | .751 | 243. | .698 | 282. | .651 | 324. | .611 | 368. | .542 | 466. | .488 | 576. |
| 4.6 | .815 | 254. | .757 | 295. | .707 | 339. | .662 | 385. | .588 | 488. | .529 | 602. |
| 4.8 | .881 | 265. | .818 | 308. | .763 | 353. | .715 | 402. | .636 | 509. | .572 | 628. |
| 5.0 | .949 | 276. | .881 | 321. | .822 | 368. | .770 | 419. | .685 | 530. | .617 | 654. |
| 5.2 | 1.020 | 287. | .947 | 333. | .883 | 383. | .828 | 435. | .736 | 551. | .662 | 680. |
| 5.4 | 1.092 | 298. | 1.014 | 346. | .947 | 397. | .888 | 452. | .788 | 572. | .710 | 707. |
| 5.6 | 1.167 | 309. | 1.083 | 359. | 1.011 | 412. | .949 | 469. | .843 | 594. | .758 | 733. |
| 5.8 | 1.245 | 321. | 1.155 | 372. | 1.078 | 427. | 1.011 | 486. | .899 | 615. | .809 | 759. |
| 6.0 | 1.325 | 332. | 1.229 | 385. | 1.148 | 442. | 1.076 | 502. | .957 | 636. | .861 | 785. |
| 7.0 | 1.75 | 387. | 1.63 | 449. | 1.52 | 515. | 1.43 | 586. | 1.27 | 742. | 1.143 | 916. |

INSIDE DIAMETER OF PIPE IN INCHES.

| | 22 | | 24 | | 26 | | 28 | | 30 | | 36 | |
|---|---|---|---|---|---|---|---|---|---|---|---|---|
| Velo. in ft. per sec. | Loss of head in feet. | Cubic feet per min. | Loss of head in feet. | Cubic feet per min. | Loss of head in feet. | Cubic feet per min. | Loss of head in feet. | Cubic feet per min. | Loss of head in feet. | Cubic feet per min. | Loss of head in feet. | Cubic feet per min. |
| 2.0 | .108 | 316. | .098 | 377. | .091 | 442. | .084 | 513. | .079 | 589. | .066 | 848. |
| 2.2 | .127 | 348. | .116 | 414. | .108 | 486. | .099 | 564. | .093 | 648. | .078 | 933. |
| 2.4 | .149 | 380. | .136 | 452. | .126 | 531. | .116 | 616. | .109 | 707. | .091 | 1018. |
| 2.6 | .171 | 412. | .157 | 490. | .145 | 575. | .134 | 667. | .126 | 766. | .104 | 1100. |
| 2.8 | .195 | 443. | .180 | 528. | .165 | 619. | .153 | 718. | .144 | 824. | .119 | 1188. |
| 3.0 | .222 | 475. | .204 | 565. | .188 | 663. | .174 | 770. | .163 | 883. | .135 | 1273. |
| 3.2 | .249 | 507. | .229 | 603. | .211 | 708. | .195 | 821. | .182 | 942. | .152 | 1357. |
| 3.4 | .278 | 538. | .255 | 641. | .235 | 752. | .218 | 872. | .204 | 1001. | .169 | 1442. |
| 3.6 | .308 | 570. | .283 | 678. | .261 | 796. | .242 | 923. | .226 | 1060. | .188 | 1527. |
| 3.8 | .340 | 601. | .312 | 716. | .288 | 840. | .267 | 974. | .249 | 1119. | .207 | 1612. |
| 4.0 | .373 | 633. | .342 | 754. | .315 | 885. | .293 | 1026. | .273 | 1178. | .228 | 1697. |
| 4.2 | .408 | 665. | .374 | 791. | .345 | 929. | .320 | 1077. | .299 | 1237. | .249 | 1782. |
| 4.4 | .444 | 697. | .407 | 829. | .375 | 973. | .348 | 1129. | .325 | 1296. | .271 | 1866. |
| 4.6 | .482 | 728. | .441 | 867. | .407 | 1017. | .378 | 1180. | .353 | 1355. | .294 | 1951. |
| 4.8 | .521 | 760. | .476 | 905. | .440 | 1062. | .409 | 1231. | .381 | 1414. | .318 | 2036. |
| 5.0 | .561 | 792. | .513 | 942. | .474 | 1106. | .440 | 1283. | .411 | 1472. | .342 | 2121. |
| 5.2 | .602 | 823. | .552 | 980. | .510 | 1150. | .473 | 1334. | .441 | 1531. | .368 | 2206. |
| 5.4 | .645 | 855. | .591 | 1018. | .546 | 1194. | .507 | 1385. | .473 | 1590. | .394 | 2291. |
| 5.6 | .690 | 887. | .632 | 1055. | .583 | 1239. | .542 | 1437. | .506 | 1649. | .421 | 2376. |
| 5.8 | .735 | 918. | .674 | 1093. | .622 | 1283. | .578 | 1488. | .540 | 1708. | .450 | 2460. |
| 6.0 | .782 | 950. | .717 | 1131. | .662 | 1327. | .615 | 1539. | .574 | 1767. | .479 | 2545. |
| 7.0 | 1.040 | 1109. | .953 | 1319. | .879 | 1548. | .817 | 1796. | .762 | 2061. | .636 | 2968. |

The following formula, deduced by Wm. Cox, gives practically the same results as the above table and will be found useful in many instances. $F = \frac{L}{1000\,D}(4V^2 + 5V - 2)$. Where F = friction head, L = length of pipe in feet; D = diameter of pipe in inches; V = velocity in feet per second.

# RIVETED SHEET STEEL PIPE.

In no other part of the world have the various systems of fluming, ditching and piping water been developed to such an extent as on the Pacific Coast, and nowhere else are the advantages of these methods of distribution so well understood. It is probably quite within bounds to say that there are anywhere from six to seven thousand miles of canals and flumes on the Pacific Coast, and perhaps quite as large an amount of piping, the latter of sizes running from four to sixty inches in diameter, carrying water with varying pressures, in some instances as high as 1,700 feet head.

The question of water conduit is everywhere one of so much importance and is so intimately connected with the utilization of power by the Pelton Wheel, that special consideration is given to it in this connection. The use of sheet iron pipe for hydraulic purposes and for power is strictly of California origin and though so extensively adopted in the Pacific Coast States, its advantages may be said to be very imperfectly

## UNDERSTOOD IN OTHER PARTS OF THE COUNTRY.

The general impression among engineers, who have not made this subject a special study, being, that heavy cast iron pipe, or lap weld tubing is necessary to carry any considerable pressure, or for any degree of permanency. This prejudice has arisen in part from the fact that sheet iron pipes when first used were only painted on the outside, which proved ineffectual to prevent corrosion, so that from eight to ten years generally covered the extreme limit of service of pipe so laid.

The plan of coating with a preparation of asphalt both inside and out was first adopted some thirty years ago, and since that time pipe of this character, as above intimated, has been so extensively used under every variety of condition as to pressure, diameter and thickness of iron as to afford the fullest and most reliable data. The result of this experience briefly stated is that a comparatively light iron, in pipe of moderate diameter, will stand a much higher head than is generally supposed, while with a proper adaptation to diameter and pressure, it is not only much cheaper but will meet every requirement more satisfactorily

## THAN ANY OTHER KIND OF PIPE MADE.

The method of covering with asphalt referred to affords perfect protection against corrosion, and so long as the coating is intact, makes it practically indestructible so far as all ordinary wear is concerned. Several pipe lines so laid have now been in service on this Coast more than twenty-five years, and are still in a very good state of preservation. The conditions which interfere with the best service are where the coating is worn off by abrasion in transportation, or where the pipe is subject to severe shock by the presence of air or, by a sudden closing of the gates, or where the service is intermittent, causing contraction and expansion which opens the joints and breaks the covering. With ordinary care these objections can mostly be overcome. While the primary object of coating pipe in this way is to prevent oxidization and thus insure its durability, it is incidentally an advantage in providing a smooth surface on the inside, which reduces the friction of water in its passage. For pipe of large diameter and heavy pressure, steel is now being largely adopted, having greater tensile strength than iron.

## LAYING THE PIPE LINE.

In laying pipe the shortest practicable distance is generally adopted, where the configuration of the ground will admit. In ordinary cases the pipe is laid from ditch, flume or other source of supply, on the surface of the ground at any angle necessary to conform thereto, and thus brought directly to the Wheel, no penstock or any direct verticle pressure being necessary; simply the head the line of pipe laid at any degree of inclination may give. Short turns or acute angles should be avoided, as they tend to lessen the pressure and give a shock to the pipe.

What is termed the slip joint is ordinarily used, excepting in pipe of large diameter or under very high head. In laying such pipe where the lengths come together at an angle, a lead joint should be made. This is done by putting on a sleeve, allowing a space say three-eights of an inch for running in lead. With a heavy pressure, and especially on steep grades the lengths should be wired together, lugs being put on the sections forming the joints for this purpose, and where the grade is very steep, the pipe

## SHOULD BE SECURELY ANCHORED WITH WIRE CABLE.

In laying the pipe line it is customary to commence at the Wheel, and with slip joint the lower end of each length should be wrapped with cotton drilling or burlaps to prevent leaking. Care being taken in driving the joints together not to move the gate and nozzle from their position. Some temporary bracing may be necessary to provide against this.

Where several Wheels are to be supplied from one pipe line, a branch from the main in the form of the letter Y is preferable to a right angle outlet. When taken from the main at a right angle, the tap hole should be nearly as large as the main, reducing by taper joint to the size of pipe attached to the wheel gate.

It is advised where practicable to lay the pipe in a trench, covering it with earth. Even in warm climates where this is not necessary as protection from frost, it is desirable to prevent contraction and expansion by variations of temperature, as well as to afford security against accident. When laid over a rocky surface, a covering of straw or manure will protect it from the sun, and generally prevent freezing, as where kept in motion, water under pressure will stand a great degree of cold without giving trouble in this way. After connections are made it should be tested before covering to see that the joints are tight.

# RIVETED SHEET STEEL PIPE.

Care should be taken when the pipes are first filled to see that the air is entirely expelled, the use of air valves being necessary in long lines laid over undulating surfaces. Care should also be taken before starting to see that there are no obstructions in the pipe or connections to Wheel, and that there are no leaks to reduce the pressure. Pipe lines of any considerable length should be graduated as to size, being larger near the top and reduced toward the lower end, the thickness of iron for various sizes being determined by the pressure it is to carry. This is a saving in first cost and facilitates transportation by admitting of lengths being run inside of each other.

The loss of head by friction in pipe depends upon their diameter, length and quantity of water passed. Enough head should be added to that given in tables to overcome the friction. Full information in regard to this, as also thickness of pipe suited to various heads will be found under appropriate headings. A pressure gauge at the Wheel gate should indicate the heads tabled while the water is flowing out of the nozzle. In making calculations to cover entire length of pipe line, an allowance of three inches on each length should be made on slip joint pipe for loss in driving the joints.

## RELATING TO SHIPMENT.

When used near railroad stations, pipe is generally made in 27 ft. lengths for purpose of economizing freight, this being the length of a car. When transported long distances by wagon, it is usually made in about 20 ft. lengths, this being a matter, however, suited to the convenience of parties ordering. For pipe of large diameter or for transportation over long distances, as also for mule packing, it is made in sections or joints of 24 to 30 inches in length, rolled and punched, with rivets furnished to put together on the ground where laid. Pipe of this character being cold riveted, is easily put together with the ordinary tools for the purpose. In such case preparation should be made for coating with asphalt before laying, material for this properly prepared, being furnished when desired in barrels or cans, as may be most convenient for transportation. Also wrought iron tanks furnished for dipping, with full instructions for use.

In dealing with pipe of large diameter a mild quality of steel is recommended, of such thickness and tensile strength as may be necessary to stand the required pressure. Material of this character is made to order in plates of any requisite dimension, cut punched, scarfed and rolled, ready for riveting up and deliverable at any point affording the best freight rates to destination.

Pipe lines of this character generally involve a large outlay and an exacting service. Consequently careful calculations as to capacity and strength, as well as in spacing and punching, are considerations of the first importance. All of this work is done under the immediate supervision of one of our engineers, and has such care and personal attention as to entitle us to the confidence of parties requiring such material.

NOTE.—In many cases much expense may be saved in pipe by conveying the water in a flume or ditch along the hillside, covering in this way a large part of the distance, then piping it down to the power station by a short line. This is more especially applicable to large plants, where the cost of the pipe is an important item.

# SECTION OF HYDRAULIC SLIP JOINT PIPE.

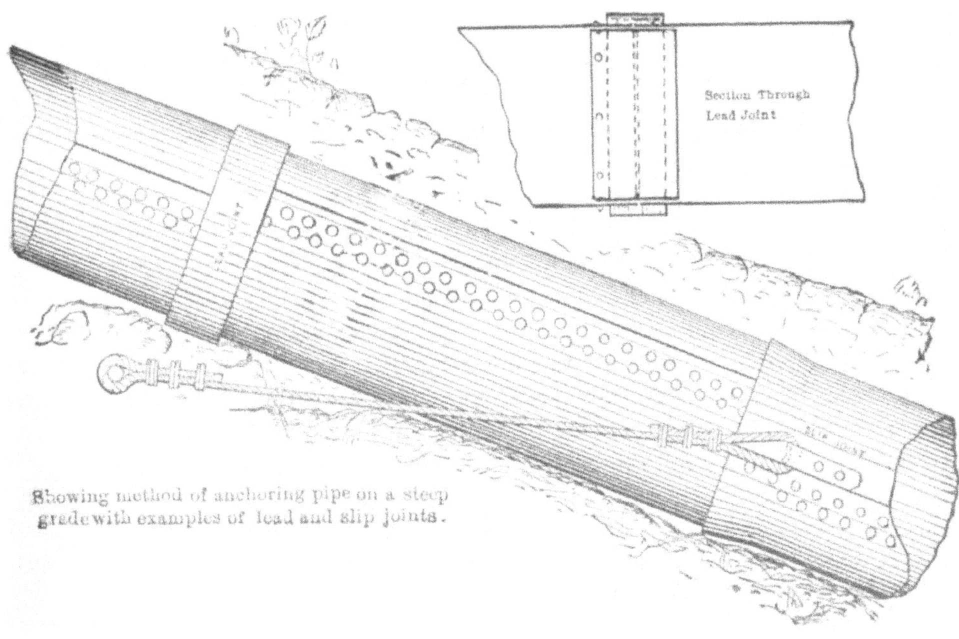


Showing method of anchoring pipe on a steep grade with examples of lead and slip joints.

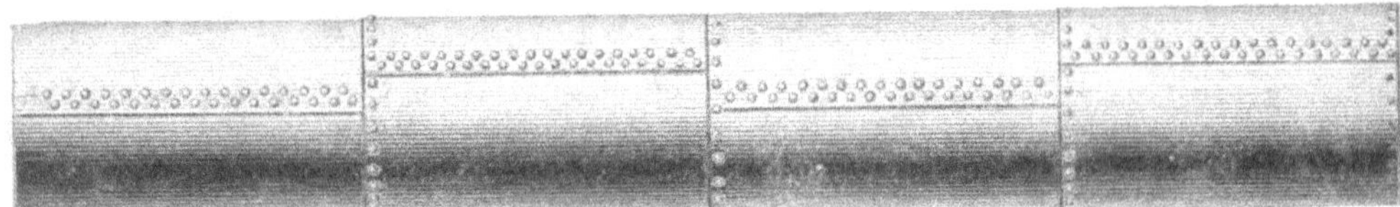

SECTION OF HYDRAULIC SLIP JOINT PIPE.

## EXPLANATION OF WEIR DAM MEASUREMENT.

Place a board or plank in the stream, as shown in the drawing, at some point where a pond will form above. The length of the notch in the dam should be from two to four times its depth for small quantities and longer for large quantities. The edges of the notch should be bevelled toward the intake side as shown. The overfall below the notch should not be less than twice its depth, that is 12 inches if the notch is 6 inches deep, and so on.

In the pond about 6 feet above the dam drive a stake, and then obstruct the water until it rises precisely to the bottom of the notch and mark the stake at this level. Then complete the dam so as to cause all the water to flow through the notch and after time for the water to settle, mark the stake again for this new level. If preferred the stake can be driven with its top precisely level with the bottom of the notch and the depth of the water be measured with a rule after the water is flowing free, but the marks are preferable in most cases. The stake can then be withdrawn and the distance between the marks is the theoretical depth of flow corresponding to the quantities in the table

## TABLE FOR WEIR MEASUREMENT.

Giving Cubic Feet of Water per minute, that will flow over a Weir one inch wide and from ⅛ to 20⅞ inches deep.

| INCHES. | | 1/8 | 1/4 | 3/8 | 1/2 | 5/8 | 3/4 | 7/8 |
|---|---|---|---|---|---|---|---|---|
| **0** | .00 | .01 | .05 | .09 | .14 | .19 | .26 | .32 |
| **1** | .40 | .47 | .55 | .64 | .73 | .82 | .92 | 1.02 |
| **2** | 1.13 | 1.23 | 1.35 | 1.46 | 1.58 | 1.70 | 1.82 | 1.95 |
| **3** | 2.07 | 2.21 | 2.34 | 2.48 | 2.61 | 2.76 | 2.90 | 3.05 |
| **4** | 3.20 | 3.35 | 3.50 | 3.66 | 3.81 | 3.97 | 4.14 | 4.30 |
| **5** | 4.47 | 4.64 | 4.81 | 4.98 | 5.15 | 5.33 | 5.51 | 5.69 |
| **6** | 5.87 | 6.06 | 6.25 | 6.44 | 6.62 | 6.82 | 7.01 | 7.21 |
| **7** | 7.40 | 7.60 | 7.80 | 8.01 | 8.21 | 8.42 | 8.63 | 8.83 |
| **8** | 9.05 | 9.26 | 9.47 | 9.69 | 9.91 | 10.13 | 10.35 | 10.57 |
| **9** | 10.80 | 11.02 | 11.25 | 11.48 | 11.71 | 11.94 | 12.17 | 12.41 |
| **10** | 12.64 | 12.88 | 13.12 | 13.36 | 13.60 | 13.85 | 14.09 | 14.34 |
| **11** | 14.59 | 14.84 | 15.09 | 15.34 | 15.59 | 15.85 | 16.11 | 16.36 |
| **12** | 16.62 | 16.88 | 17.15 | 17.41 | 17.67 | 17.94 | 18.21 | 18.47 |
| **13** | 18.74 | 19.01 | 19.29 | 19.56 | 19.84 | 20.11 | 20.39 | 20.67 |
| **14** | 20.95 | 21.23 | 21.51 | 21.80 | 22.08 | 22.37 | 22.65 | 22.94 |
| **15** | 23.23 | 23.52 | 23.82 | 24.11 | 24.40 | 24.70 | 25.00 | 25.30 |
| **16** | 25.60 | 25.90 | 26.20 | 26.50 | 26.80 | 27.11 | 27.42 | 27.72 |
| **17** | 28.03 | 28.34 | 28.65 | 28.97 | 29.28 | 29.59 | 29.91 | 30.22 |
| **18** | 30.54 | 30.86 | 31.18 | 31.50 | 31.82 | 32.15 | 32.47 | 32.80 |
| **19** | 33.12 | 33.45 | 33.78 | 34.11 | 34.44 | 34.77 | 35.10 | 35.44 |
| **20** | 35.77 | 36.11 | 36.45 | 36.78 | 37.12 | 37.46 | 37.80 | 38.15 |

**Example Showing the Application of the above Table.**

Suppose the Weir to be 66 inches long, and the depth of water on it to be 11⅝ inches. Follow down the left left hand column of the figures in the table until you come to 11 inches. Then run across the table on a line with the 11, until under ⅝ on top line and you will find 15.85. This multiplied by 66, the length of Weir, gives 1046.10, the number of cubic feet of water passing per minute.

## EXPLANATION OF MINER'S-INCH MEASUREMENT.

The term MINER'S INCH is of California origin, and not known or used in any other locality, it being a method of measurement adopted by the various ditch companies in disposing of water to their customers. The term is more or less indefinite, for the reason that the water companies do not all use the same head above the center of the aperture, and the inch varies from 1.36 to 1.73 cubic feet per minute each, but the most common measurement is through an aperture 2 inches high and whatever length is required, and through a plank 1¼ inches thick, as shown in print. The lower edge of the aperture should be 2 inches above the bottom of the measuring box, and the plank 5 inches high above the aperture, thus making a 6-inch head above the center of the stream. Each square inch of this opening represents a miner's inch, which is equal to a flow of 1½ cubic feet per minute.

THE PELTON TABLES are all based upon the measurement above given.

Time is not to be considered in any calculation based upon miner's inches.

**MEASUREMENT IN AN OPEN STREAM BY VELOCITY AND CROSS SECTION.**

Measure the depth of the water at from 6 to 12 points across the stream at equal distances between. Add all the depths in feet together and divide by the number of measurements made; this will be the average depth of the stream, which multiplied by its width will give its area or cross section. Multiply this by the velocity of the stream in feet per minute, and you will have the cubic feet per minute of the stream.

The velocity of the stream can be found by laying off 100 ft. on the bank and throwing a float into it at the middle, noting the time passing over the 100 ft. Do this a number of times and take the average. Then dividing this distance by the time gives the velocity in feet per minute at the surface. As the top of the stream flows faster than the bottom or sides,—the difference being about 8 per cent—it is better to measure a distance of 120 ft. for float and reckon it as 100.

# ADVANTAGES OF THE PELTON SYSTEM.

BY ROSS E. BROWNE, MINING AND HYDRAULIC ENGINEER.

The function of a water wheel, operated by a jet of water escaping from a nozzle, is to convert the energy of the jet, due to its velocity, into useful work. In order to utilize this energy fully, the wheel bucket, after catching the jet, must bring it to rest before discharging it, without inducing turbulence or agitation of the particles. It is plain that this cannot be fully effected, and that unavoidable difficulties necessitate the loss of a portion of the energy. The principal losses occur as follows:

First: In sharp or angular diversion of the jet in entering, or in its course through the bucket, causing impact, or the conversion of a portion of the energy into heat instead of useful work.

Second: In the so-called frictional resistance offered to the motion of the water by the wetted surfaces of the buckets, causing also the conversion of a portion of the energy into heat instead of useful work.

Third: In the velocity of the water, as it leaves the bucket, representing energy which has not been converted into work.

Hence, in seeking a high efficiency, there are presented the following considerations:

1st. The bucket surface at the entrance should be approximately parallel to the relative course of the jet, and the bucket should be curved in such a manner as to avoid sharp angular deflection of the stream. If, for an example, a jet strikes a surface at an angle and is sharply deflected, a portion of the water is backed, the smoothness of the stream is disturbed, and there results considerable loss by impact and otherwise. The entrance and deflection in the PELTON bucket are such as to avoid these losses in the main.

2d. The number of buckets should be small, and the path of the jet in the bucket short; in other words, the total wetted surface should be small, as the loss by friction will be proportional to this.

A small number of buckets is made possible by applying the jet tangentially to the periphery of the wheel, as provided in the construction of the PELTON.

3d. The discharge end of the bucket should be as nearly tangential to the wheel-periphery, as compatible with the clearance of the bucket which follows; and great differences of velocity in the parts of the escaping water should be avoided. In order to bring the water to rest at the discharge end of the bucket, it is easily shown, mathematically, that the velocity of the bucket should be one-half the velocity of the jet.

An ordinary curved or cup bucket will cause the heaping of more or less dead or turbulent water in the bottom of the bucket. This dead water is subsequently thrown from the wheel with considerable velocity, and represents a large loss of energy.

The introduction of the wedge in the PELTON bucket is an efficient means of avoiding this loss.

A wheel of the form of the PELTON conforms closely in construction to each of these requirements. The entrance and deflection of the jet is smooth and induces very little shock. The discharge from the wheel, when running at proper speed, is almost entirely at the bottom of the wheel, and the amount of water carried over is small, *i. e.* the loss due to the energy of the discharged water is unusually small.

Both feed and discharge are practically tangential. The tangential feed admits of the design with a minimum number of buckets—the bucket is short—in fact, the entire wetted surface is small, and an important advantage is gained over the partial turbines with inner feed. The bucket is open and the rapid passage of the jet is comparatively free.

In the tests of the PELTON WHEEL made by me at the university of California, the diameter of the wheel was 15 inches, the width of the bucket 1.5 inch, and the efficiencies shown under 50 foot head were as follows: With a 7-16 inch nozzle, 82.6 per cent. With a ⅜ inch nozzle, 82.5 per cent.

The efficiency was determined under as low a head as 8 feet, still showing 73 per cent.

It is proper to state that the wheel with which the above tests were made was constructed in the workshop of the university and did not conform wholly to the company's standard. The size of the bucket was too small to do full justice to the wheel, owing to the difficulty of shaping the curves accurately. It is claimed that tests with larger wheels have given higher efficiencies, and I have no reason for doubting the claim.

In connection with the university experiments referred to, comparative tests were also made with a partial turbine of the Girard type and the following conclusion reached: The PELTON WHEEL, beside giving a higher efficiency, is simpler in construction, has a decided advantage in the setting of the nozzle, and is not so dependent upon the precise size of nozzle used.

I do not hesitate to express the following opinions regarding the PELTON WHEEL:

1st. It will give a high efficiency under a wide range of heads, say from 20 feet upward.

2d. It will be equally efficient whether operated by very small or very large nozzles, provided the proper ratio is maintained between the diameter of nozzle and size of bucket, etc.

3d. Its general simplicity of construction is a matter of great advantage in its application as a motor. In designing a plant, the size or number of wheels can be more readily adapted to the requirements of speed of shafting and distribution of power, without sacrifice of efficiency or entailment of extravagant cost. It becomes practicable to make more direct applications of the power, frequently enabling the avoidance of counter-shafting, etc.

4th. It meets the requirements of a high-head wheel much more efficiently than do the common forms of turbine. It is evidently far better adapted to high-heads than the closed or full-running turbines.

Closed turbines are distinctly LOW HEAD wheels, and are not efficient motors under high heads. A high velocity of the enclosed wheel of a full-running turbine causes great agitation and turbulence of the confined water and it is readily apparent that there occur extravagant losses by impact. The superior efficiency of the PELTON, as a high-head motor, is due to the high efficiency of the circular nozzle, the smooth and rapid deflection of the water in passing through the open bucket, and the small aggregate amount of wetted surface in the buckets.

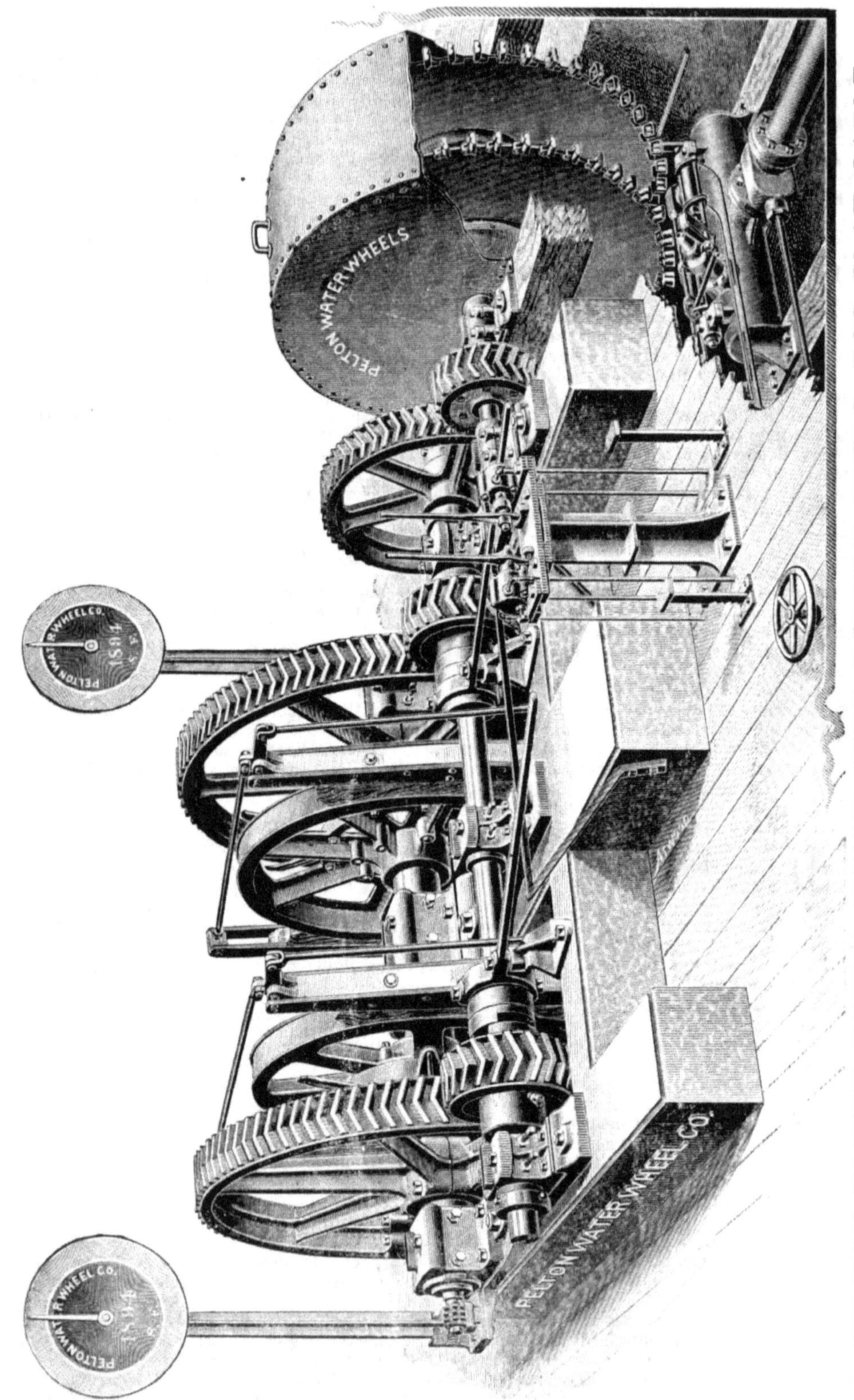


PELTON WATER WHEEL HOIST WITH DOUBLE REELS, FOR FLAT CABLE.

## PELTON WATER-POWER HOIST.

The illustration on the opposite page shows a water-power hoist of recent and modern construction, embracing all known safety appliances and improvements in this class of machinery. This style of hoist has been generally adopted on the Pacific Coast where water-power is available, and embodies the results of many years of practical experience in work of this character.

The design presented is a flat rope double reel hoist for a two-compartment shaft recently built by the PELTON WATER WHEEL Co. for the Milwaukee mine, in the Cœur d'Alene District, Idaho.

The wheels are 6 feet diameter, made of solid steel discs with phosphor bronze buckets, and run under 850 feet head at a speed of 375 revolutions, which gives the wheels a peripheral velocity of 6,800 feet per minute. Water is applied to the wheels through nozzles, the gates of which are controlled by hydraulic valves operated by levers at the engineer's stand.

An air receiver is located at the lower end of the pipe to prevent shock to the line by a sudden closing of the gate. The nozzles are provided with deflectors operated by levers, which give the engineer control of the wheels, independent of any movement of the gates, admitting of their being stopped instantly in case of accident.

The hoist is equipped with powerful post brakes operated by foot treadles, which afford absolute security in stopping or lowering the cages, and give perfect control of the movement at all points in the shaft, while, to afford additional security, the wheels may be instantly reversed without shock or injury, by applying the stream in an opposite direction.

The reels are mounted on separate shafts and connected by gearing and clutches on pinion shaft, so that they may be operated together or independently, as may be desired. Dial indicators operated by worm gears on reel hubs are attached to each reel, showing position of cage in shaft.

The hoist here referred to carries a load of 5,500 pounds at a speed of 400 feet per minute, and is designed to work to a depth of 1,000 to 1,500 feet. The gearing can be arranged to increase the speed to 600 feet per minute if desired, and when operated under a lower head of water the intermediate gears can be dispensed with.

A ¾-inch stream under head named gives each wheel a capacity of 70 horse-power, and a 1-inch stream 120 horse-power. Similar hoists with reels for both flat and round rope have been furnished by the PELTON COMPANY to many of the most prominent mines on the coast, some working to a depth of more than 2,000 feet.

The cost of operating and of maintenance of a plant of this character is merely nominal as compared with steam, while by the various means indicated it is under much better control, as, having no fly wheels, the speed can be checked much quicker in case of emergency. The wheels and appliances connected therewith, constituting the power part of the plant above described, weigh only 2,800 pounds, while steam-power of the same capacity would weigh not less than 30,000 pounds, thus indicating the great advantage of water-power where by any reasonable means it can be made available.

## THE FRANKLIN INSTITUTE AWARD.

THE PELTON WATER WHEEL COMPANY have recently had a distinguished honor paid them in the award of the Franklin Institute, of Pennsylvania, of the ELLIOT CRESSON gold medal, made as a just recognition of what is regarded by that institution as one of the most useful and illustrious inventions of the age.

This award is accompanied by an elaborate and exhaustive report made by a committee of scientists appointed for the purpose, which traces the history and progress of water-wheel practice from the early centuries down to the most modern and improved methods. The report concludes with the following paragraph:—

"We have given this invention the fullest and most unbiased examination permissible, within the time allowed us and within the compass of the literature at our command. We conclude that the PELTON WATER WHEEL possesses all the advantages of simplicity of construction, economy of installation and maintenance, adaptability to extreme heads of water, transportability, a close and sensitive automatic regulation, and of high speed, which belong to other wheels of its class which have preceded it, but FOR EFFICIENCY IT HAS EXCELLED ALL OTHERS.

"We therefore deem the PELTON WATER WHEEL worthy of the ELLIOT CRESSON MEDAL, and hereby recommend the award of same to LESTER A. PELTON, the inventor of this wheel."

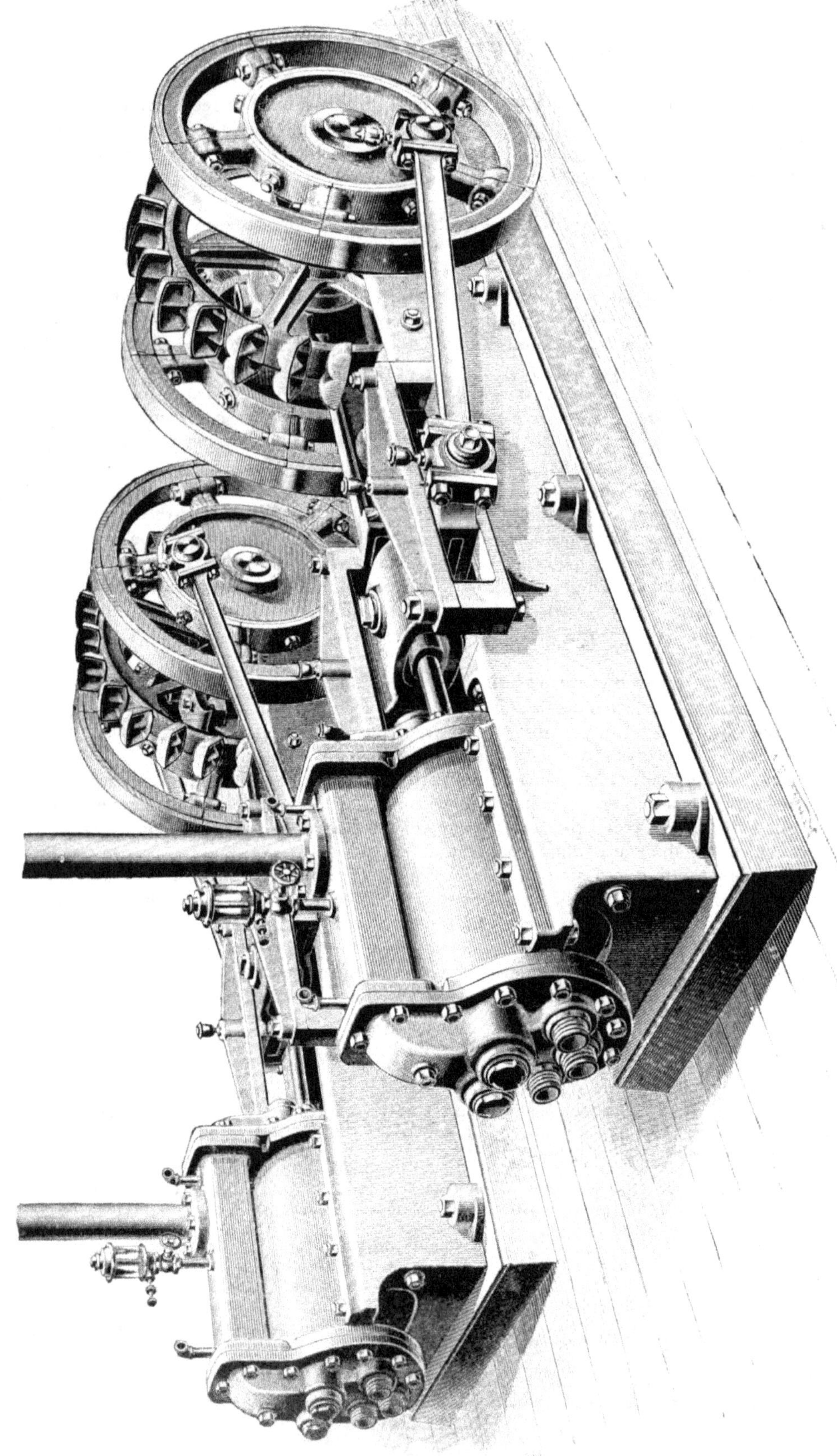

PAIR OF INGERSOLL-SERGEANT COMPRESSORS WITH PELTON WHEELS.

## AIR COMPRESSORS AND PELTON WHEELS.

The illustration on the preceding page shows a pair of Ingersoll-Sergeant compressors driven by PELTON WHEELS attached to driving shafts direct.

This plant was constructed by the Ingersoll-Sergeant Company for a mine in Peru, S. A., located in the heart of the Andes, some 100 miles east of Lima, and was made in sections for mule-packing, no piece in the entire outfit exceeding 300 pounds in weight.

The wheels are 4 feet diameter and run under a head of 50 feet, making 120 revolutions. Water is brought from a mountain stream through a line of 18-inch pipe, and is applied to the wheels by double nozzles with 3-inch openings. The air cylinders are 10 inches diameter by 18-inch stroke, and the wheels under this comparatively low head afford sufficient power to maintain 70 pounds pressure on the receiver at an elevation of 10,000 feet above sea level.

The advantage of attaching the wheels to the compressor shafts are: great economy of power, as well as in first cost, as also lessened cost of freight—a matter of first importance in machinery destined to such remote and inaccessible localities. The advantages so simple and efficient a connection afford are also manifest in the lessened cost of setting up the machinery, as well as in the absence of belt connections, the maintenance of which with loss of power involved are matters of the greatest importance in many mining localities, where every pound of water must be made to do its utmost duty.

This application of the PELTON WHEEL is made with equal facility to all forms of compressors, as well as blowers and many other classes of machinery, and where the head admits, the wheel can be made heavy enough to serve as a fly wheel, thus simplifying and cheapening still more this manner of applying power. Wheels in such cases may be made of any size, ranging from 6 up to 12 or 16 feet in diameter, as may be necessary to give proper speed to the machinery they are designed to run, suiting the buckets and nozzle delivery to conditions as to head and power requirement.

Fly wheels were made necessary in the case here referred to from the fact that the water head was too low to admit of making the wheels large enough to carry the weight required for combined motor and fly wheel and give the necessary speed to compressor.

## DUPLEX COMPRESSOR WITH PELTON WHEEL.

The print on the following page shows a 14x22 Duplex Rand Compressor driven by a PELTON WHEEL 14 feet 6 inches in diameter attached to the shaft direct. This compressor was constructed by the Rand Drill Company for the La Union Mine, Costa Rica, C. A., and runs at a speed of 95 revolutions, with a capacity for twelve No. 2 drills, requiring 120 horsepower. Weight, exclusive of wheel, 13,500 pounds. The wheel weighs 5,000 pounds, and, as will be observed, not only furnishes the motive power, but is sufficiently heavy to serve as a balance wheel; being made in sections, it was easily transported and attached to shaft. It runs under a head of 380 feet, developing the power above mentioned, with ample margin for contingencies.

The wheel was made of this large diameter to give proper speed to the compressor under head of water available, the buckets and nozzle being of a size to develop the required power with the greatest degree of economy and efficiency. It will be evident that the wheel may be made of any size to conform to conditions as to head of water and speed of compressor. Also of weight to answer as fly-wheel, where the head of water will admit of making it of a diameter not less than about 6 feet.

Attention is invited to the following statement from the superintendent of the above company regarding the operations of this plant.

"The large PELTON WHEEL on our compressor shaft, which answers for both fly wheel and prime mover, has never given us any trouble during the three years it has been in constant service, and gives entire satisfaction. There can be no two opinions about the advantages of such a power or the means of applying it. Nothing can be more simple and reliable or more efficient in performance. The 5-foot PELTON driving our quartz mill, and the 3-foot driving the sawmill, give equally good results. The latter is mounted directly upon the saw arbor—as is the case with the compressor wheel—doing away with belts and counter shaft and consequent friction and strains, making a mill of the utmost simplicity and power.

"The high efficiency of these wheels, under both high and low heads, is a matter of great importance in all localities, but more especially in a country like this, where we want to get the utmost power that varying quantities of water can give. I may also add that their absolute reliability, under all the varying conditions of use, are considerations of great moment, making the most convenient, economical and satisfactory power that can be conceived of."

NOTE.—A large number of additional references could be given of wheels running compressors as well as pumps, blowers, and other classes of machinery by direct connection to crank shaft, as herein shown, but those noted will serve to show the practicability and advantage of this application where the conditions as to head of water and speed of machinery favor it.

The PELTON—it may be here stated—is the only wheel that admits of such direct connection with any high efficiency or satisfactory service. The saving in first cost and power as well as the general favorable results attending such application, have given it the preference for all this class of work.

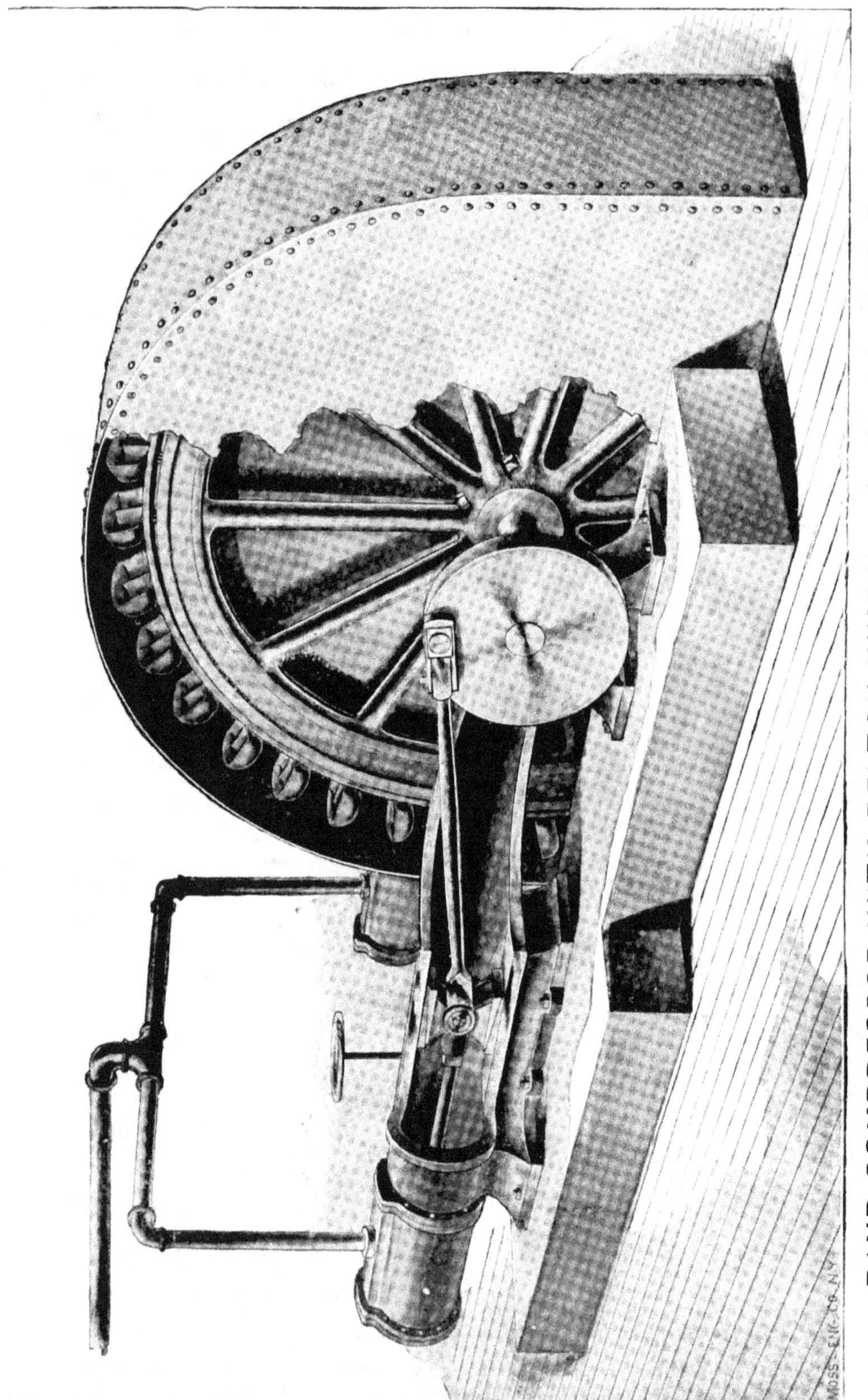

RAND COMPRESSOR WITH PELTON WHEEL DIRECT CONNECTED.

DUPLEX RAND COMPRESSOR WITH PELTON WHEEL.

## AIR COMPRESSION BY WATER POWER.

Extract from a paper on the Transmission of Power by Compressed Air, having especial reference to the installation of an extensive plant of this character at Niagara Falls—by William L. Saunders, Mem. Am. Soc. C. E.

"The remarkable results that have been obtained in practical experience with PELTON WHEELS have materially added to the economy of compressing air by water-power.

"The PELTON is unquestionably the simplest, and my experience with it justifies the statement that it is the most reliable and economical water wheel in use. It is especially adapted to the peculiar conditions existing at Niagara Falls, in that the head is high and the volume of water large.

"I have given this subject much study, and have made some plans and suggestions, which have been referred to the Consulting Engineers of the Niagara Falls Commission. According to my idea, the simplest and best plan is one in which one or more PELTON WHEELS are driven at the bottom of the shaft, that is, at the point of the greatest head. The power of the water is then converted into reciprocating motion by a Pitman standing vertically, and properly balanced, working air-compressing cylinders through bell cranks. Such an application as this is free from gears, and has really only one shaft, that of the PELTON WHEEL.

"The Turbine Wheel, which has been used for so many years for this purpose, is expensive in cost of plant, expensive in repairs, and wasteful of water, while the PELTON is extremely simple, and where the size of the wheel is in proportion to the head of water, with the buckets properly constructed, the value of the water-power —that is, the fall or head, multiplied by the volume, becomes transformed into the inertia of a fly wheel, to the shaft of which is attached suitable means for reciprocating the piston of an air cylinder. This fly wheel is large enough and heavy enough to store the power of the water and to give it out in the shape of work in an economical way.

"The loss is, in the first place, a certain amount of friction in the water pipe, wheel, and buckets, a certain amount of friction in the compressor, and a certain loss by heat, leakage, and clearance in the air cylinder. My experience has taught me that these losses by means indicated have been reduced to a very low figure.

"Obviously, the greatest amount of power that is obtained in a waterfall is the product of the volume, that is, the weight per cubic foot multiplied by the head or fall. With the PELTON WHEEL the full value of this water-power is given at the nozzle, less only a little friction in the pipe and nozzle. This force, which is represented by the jet of water, is intercepted and converted into mechanical movement; and in doing this, if no water is lost, and the movement of the water is transferred into the movement of the wheel, there is the nearest approach to the perfect utilization of water-power.

"As a matter of fact, the leakage is a very small figure, and when the speed of the wheel is properly proportioned to the spouting velocity of the water, the buckets simply take the energy out of the stream and leave the water inert under the wheel. After converting the power of the water into the revolving movement of the wheel, it is a simple matter to reciprocate pistons of air compressors; and for such a plant as proposed, a large number of air cylinders may be arranged in rows and operated by a single upright Pitman properly balanced."

"The application here suggested is one that has been made by the PELTON WATER WHEEL COMPANY in a large number of instances to both air and ammonia compressors, as well as to electric generators, pulp grinders, ventilating fans, pumps, and various other classes of machinery. The many advantages of such a connection without intermediate gearing, is one of the important and distinctive features of the PELTON SYSTEM OF POWER, which has served to bring it into such prominence and general use in all parts of the industrial world."

## PELTON WHEEL OPERATING AN ICE MACHINE

The above cut shows a PELTON WHEEL driving a Dow Refrigerating Machine. The wheel, as will be observed, is attached to the shaft direct and made to serve the double purpose of prime mover and balance wheel. It is 9 feet in diameter, and runs under a head of 60 feet, with a speed of 65 revolutions per minute. The nozzle and gate are in the rear, and consequently can not be shown in the view given. Two of these plants have been operated by the Azusa Ice Co. in Southern California for the past three years, turning out 30 tons of ice per day. The wheels have shown a high degree of efficiency, and have run continuously during the time named without any repairs whatever. The same application, it is evident, can be made to any other form of compressor, either vertical or horizontal, as also of a diameter and capacity to suit any requirement as to speed and power.

## PNEUMATIC POWER TRANSMISSION.

AN EIGHTEEN-FOOT PELTON WHEEL

DRIVING A DUPLEX TANDEM AIR COMPRESSOR.

## A STUPENDOUS COLUMN OF WATER.

Probably the most extraordinary water-power installation, so far as head is concerned, that can be referred to, is that made some four years ago in one of the famous Comstock mines, at Virginia, Nevada.

This consisted of a 36-inch PELTON WHEEL, made of a solid steel disc, with phosphor bronze buckets securely riveted to the rim, and is located at the Sutro Tunnel level of the California and Consolidated Virginia shaft, 1,640 feet below the surface. In addition to the head afforded by the depth of the shaft, the pipe is connected to the line of the Gold Hill Water Company, which carries a head of 460 feet, giving the wheel a vertical head of 2,100 feet, equivalent to a pressure of 911 pounds. The water, after passing over the wheel, is carried out through the tunnel, 4 miles in length. The wheel runs at 1,150 revolutions, which gives it a peripheral velocity of 10,804 feet per minute, or about 120 miles an hour.

The construction of the wheel amply provides for the centrifugal strain the velocity of the water gives it, running without load, when it would attain the enormous speed of 21,608 feet per minute, equal to about 240 miles per hour. A nozzle tip ½ inch in diameter gives under above conditions 100 horse-power. Every miner's inch of water, equal to a flow of 1.6 cubic feet per minute, gives 5 horse-power, while 1 horse-power is given for every 2 pounds of metal in the wheel. It is only by comparison that an idea can be obtained of the height of a column of water due to such pressure. It is more than four times as high as the Washington Monument, and considerably more than twice the height of the Eiffel Tower.

Though running under such an extreme head, no repairs have been required, and a high degree of efficiency has been uniformly maintained.

The Hydraulic Power Co., of Chester, England, have a large number of PELTON WHEELS of similar construction operating in connection with various high-pressure power systems. All of these run under no less a pressure than 700 pounds, and some of them under 1,000 pounds, the latter equal to a head of 2,310 feet, wheels 18 inches in diameter, weighing but thirty pounds, giving under this head upwards of 21 horse-power with ¼-inch nozzle tips. It is evident from these examples that the head under which these wheels can be safely and efficiently operated is only limited by the strength of material in pipe and connections.

## THE NORTH STAR POWER PLANT.

The paper read by Mr. Arthur de Wint Foote, at the meeting of the American Society of Civil Engineers, which convened in San Francisco, two years ago was of uncommon interest and the subject of much discussion.

The paper referred to was a very elaborate and exhaustive report upon the operation of the compressed air transmission plant of the North Star mine in Grass Valley, Cal., a brief description of which, compiled from this report, we here present.

The power station consists of a PELTON WHEEL 18 feet 6 inches in diameter, attached to the shaft of a Rix Duplex Air compressor—compound tandem type. The initial cylinders are 18 inches and the second cylinders 10 inches in diameter with a 24-inch stroke.

The wheel is built up of angle iron plates riveted together to break joints, and is held concentric with the shaft with 12 pairs of radial spokes of 1¼-inch rod iron secured by nuts to the cast-iron hub. The driving force being applied to the rim, is transferred to the hub by four pairs of 2-inch rods so arranged as to form a truss.

The wheel weighs 10,500 pounds and runs at 110 revolutions under 750-foot head, developing upwards of 300 horse-power. It is made of this large diameter for purpose of giving proper speed to the compressor under the high head available. The water is applied to the wheel through a variable nozzle controlled by a hydraulic regulator which maintains a uniform speed on the wheel with a variation from full load down to 25 per cent of same, making its operation absolutely automatic, as well as economizing the water supply, no more being used at any time than required for the work.

The construction of the wheel, as will be seen from print on opposite page, forms an ingenious mechanical combination, altogether novel and without precedent, affording an ample factor of safety with a very high peripheral velocity.

The water supply is brought to the wheel through 2½ miles of 22-inch riveted pipe, affording sufficient capacity to develop 800 horse-power. A 6-inch lap-weld pipe conveys the air at a pressure of 90 pounds from the power-house to the company's shaft, 800 feet distant, and 125 foot elevation. This is at present running a 100 horse-power hoisting engine and a 75 horse-power compound pump, besides other pumps, drills, forges, etc.

The report above referred to shows that repeated tests on this wheel made by the most approved methods and checked up very closely give the remarkable efficiency of 93 per cent at full load, and an average efficiency of something over 90 per cent for ¼, ½, ¾, and full loads.

The efficiency of compression and transmission from water wheel to motors, not including cost of reheating, is given as 79 per cent, making a most favorable showing for the plant as a whole, under the conditions here installed.

The general manager of the above-named mine states that no repairs have been required during the three years this wheel has been in operation and that the same high degree of efficiency has been constantly maintained.

## A COMBINED STEAM AND WATER POWER COMPRESSOR.

The cut below illustrates a most interesting installation recently made on the Alaska-Treadwell mine, Douglass Island, Alaska. This consists of a Duplex-Riedler Compressor, with air cylinders 24 inches in diameter by 36-inch stroke, driven by a horizontal, cross, compound, condensing engine, with steam cylinders 24x38x36 inch stroke.

The steam cylinders are placed behind the air cylinders, the piston rods being connected by threaded couplings in halves, admitting of disconnecting when operating by water-power.

The compressor has a capacity when running at 75 revolutions of delivering 2,800 cubic feet of free air per minute. The wheel is 22 feet in diameter, direct connected to the shaft of the compressors—serving the purpose of a fly-wheel as well as prime mover. It weighs 25,000 lbs. and running under a head of 480 feet at a speed of 75 revolutions, develops upwards of 500 horse-power.

An hydraulic speed regulator is attached to the wheel, which controls its movements so that a uniform air pressure is maintained on the receiver at all times.

This is the largest tangential wheel ever constructed, and shows the remarkable facility with which PELTON WHEELS can be adapted to unusual and extraordinary conditions.

In a case like this, the transmission machinery to carry such an amount of power would involve a heavy outlay, as well as constant expense in maintenance, besides material loss in power. A direct connection of water wheel to machinery to be operated without intermediate gearing is, therefore, obviously of great advantage whenever possible.

The superintendent of the above company reports this plant as working very smoothly and with a high degree of efficiency, saving fully twenty per cent in power over the usual means of transmission by rope drive.

The compressor and engines were furnished by Fraser & Chalmers, of Chicago.

## GREAT CABLE INCLINE OF MOUNT LOWE RAILWAY.

The Mount Lowe Railway, running from the Valley of San Gabriel, in Southern California, to the summit of the mountain—6,000 feet above the sea—the main incline of which is here illustrated, is one of the most famous mountain roads in the world. The application of power in this installation is of so unusual a character that some details in regard to it will be of interest.

Two sources of water supply were available under widely different conditions as to head and quantity, both of which it was desired to utilize. One of the streams comes from the summit of Echo Mountain with a fall of 1,250 feet, and the other from Rubio Cañon with a fall of 287 feet, the water from each being brought down to the station in separate lines of pipe and applied to the wheels separately. The power station consists of two Pelton wheels of the steel disc type, mounted on the same shaft with iron housing and bed plate.

The high-head wheel is 40 inches in diameter and develops 100 horse-power with a 1-inch nozzle, while the wheel for the lower head is 19 inches in diameter, to which are applied two streams developing 50 horse-power, both wheels running at 800 revolutions, the varying diameters being necessary to give the same speed and still conform to the velocity due to their respective heads. A Pelton differential governor with constant speed motor is mounted on the same bed plate, which gives control of all variations of speed due to changes of load on the generator. The power developed is transmitted to the top of Echo Mountain, where it drives the hoisting machinery of the cable incline here shown, which is 3,000 feet in length and laid on a grade of 48 percent.

# ELECTRIC POWER TRANSMISSION

A great industrial revolution is taking place in the development and various applications of electric power, the full significance and far-reaching effect of which are as yet neither understood nor appreciated. All the mystery, doubt and incredulity of the past have given way to the logic of fact and actual demonstration.

Electrical energy in its present stage of development is recognized as the MOST POTENT OF ALL FORCES and a most important factor in material progress and civilization. This subtle and universally prevalent element, about which so little was known but a short time ago, is now gathered, controlled and distributed with a CERTAINTY, PRECISION AND ECONOMY ALMOST INCREDIBLE.

But a decade has passed since it was not possible to transmit power by this means in a commercial way more than a few hundred feet. Now from thirty to forty miles is quite within the limit of ordinary practice with no restrictions as to capacity, save the expenditures in conductor and plant for generating the power.

The greatest progress in the transmission of electrical energy has, however, been in connection with the utilization of water power. With this as a motive force, under all ordinary conditions, electricity is conceded to be the MOST FLEXIBLE, RELIABLE AND ECONOMICAL POWER KNOWN.

With the interest now attaching to this subject, the means of making these vast resources available for power purposes in the most simple and efficient way has come to be of supreme importance, enlisting, as it has, the highest constructive ability and engineering skill known to modern science.

THE PELTON WATER WHEEL COMPANY have demonstrated BEYOND ALL QUESTION their claim to PRE-EMINENCE AND SUPERIORITY in this department of hydraulic engineering, as evidenced by the fact that their wheels are running a majority of the stations of this character in the United States, as well as in most foreign countries.

PELTON WHEELS meet so fully the exacting requirements of this service as regards HIGH EFFICIENCY, CLOSE REGULATION, ABSOLUTE RELIABILITY and small cost of maintenance, that they have come to be regarded as the most essential part of the equipment for an electric power plant, and NO OTHER WILL BE SERIOUSLY CONSIDERED when the advantages above mentioned are understood and appreciated. The system is so flexible that it admits of adaptation to all conditions and every variety of service, and in so simple a way as to provide against the possibility of accident or any interruption to continuous service.

The advantages of this form of power in mining operations are too well known to dwell upon. A few hastily strung wires running to any point instantly transmits the energy of the waterfall into an available force readily adapted to any service. Upwards of five hundred mines in this country are now supplied to a more or less extent with electricity for power and light, by which means all the various operations of mining and milling ores are greatly simplified and cheapened. Many such enterprises may be referred to, now on profitable basis, that owe their very existence to the ECONOMIES WHICH THIS SYSTEM OFFERS, to say nothing of the facilities and conveniences afforded for extended operations.

The question of water wheel regulation, which has so long been a souree of perplexity and annoyance in operating electric power plants, has now been definitely solved, and this Company is prepared to GUARANTEE ABSOLUTE AND RELIABLE REGULATION covering the most exacting requirements of any service.

On the following pages will be found a list of electric power installations made by this company, most of them within the past three years, aggregating some 80,000 h. p. Though operating under widely different, and in some instances most extraordinary, conditions, encountering frequently almost insurmountable difficulties, all have been an engineering success, and, so far as known, remunerative in a financial way.

The longest transmissions involved in any of these references are those of the Southern California Electric Power Company—80 miles—and the San Joaquin Electric Power Company—67 miles. Several others range from 15 to 30 miles—the loss in transmission varying from 10 to 25 per cent.

## LIST OF ELECTRIC POWER INSTALLATIONS.

|  |  | | |  |
|---|---|---|---|---|
| San Joaquin Electric Power Co., Fresno, Cal...... | 2,000 h. p. | running | under | 1,400 ft. head |
| Regla Electric Power Transmission Co., Mexico. | 3,000 h. p. | " | " | 800 ft. head |
| Big Cottonwood Power Co., Salt Lake City, Utah. | 3,000 h. p. | " | " | 380 ft. head |
| Folsom Electric Power Co., Folsom, Cal............. | 4,000 h. p. | " | " | 55 ft. head |
| Nevada County Electric Power Co., Cal............. | 1,000 h. p. | " | " | 210 ft. head |
| Santa Ysabel Mine, Tuolumne Co., Cal.............. | 500 h. p. | " | " | 300 ft. head |
| Tuolumne County Electric Co., Columbia, Cal.... | 500 h. p. | " | " | 950 ft. head |
| Gold Valley Mining Co., Downieville, Cal........... | 250 h. p. | " | " | 200 ft. head |
| Boza Electric Power Co., Venezuela, S. A .......... | 1,200 h. p. | " | " | 400 ft. head |
| Big Creek Electric Power Co., Santa Cruz, Cal... | 800 h. p. | " | " | 840 ft. head |
| Redlands Electric Power Co., Redlands, Cal........ | 1,000 h. p. | " | " | 500 ft. head |
| Petropolis Electric Power Co., Brazil, S. A......... | 800 h. p. | " | " | 260 ft. head |
| Quezaltenango Electric Co., Guatemala, C. A...... | 250 h. p. | " | " | 55 ft. head |
| Ontario Mining Company, Park City, Utah......... | 300 h. p. | " | " | 120 ft. head |
| Alaska Treadwell Mine, Douglas Island, Alaska.. | 600 h. p. | " | " | 460 ft. head |
| Colorado Springs Contract Co., Colorado .......... | 440 h. p. | " | " | 600 ft. head |
| Silver Lake Mines, Silverton, Colorado............. | 700 h. p. | " | " | 180 ft. head |
| Roaring Fork Electric Power Co., Aspen, Col..... | 1,250 h. p. | " | " | 330 ft. head |
| People's Electric Light & Power Co., Aspen, Col.. | 700 h. p. | " | " | 180 ft. head |
| Telluride Electric Power Co., Telluride, Col........ | 1,000 h. p. | " | " | 500 ft. head |
| Caroline Mining Co., Ouray, Colorado................ | 400 h. p. | " | " | 500 ft. head |
| Mount Morgan Mining Co., South Africa........... | 600 h. p. | " | " | 120 ft. head |
| Hilo Electric Light Company, Hilo, H. I............ | 250 h. p. | " | " | 260 ft. head |
| Walla Walla Electric Co., Washington.............. | 750 h. p. | " | " | 60 ft. head |
| Amecameca Electric Light & Power Co., Mexico. | 700 h. p. | " | " | 980 ft. head |
| Nelson Electric Light & Power Co., Nelson, B. C. | 350 h. p. | " | " | 160 ft. head |
| Juneau Elec. Light & Power Co., Juneau, Alaska. | 200 h. p. | " | " | 108 ft. head |
| Bucaramanga Electric Light Co.,Colombia, S. A. | 400 h. p. | " | " | 53 ft. head |
| Kyoto Electric Power Co., Kyoto, Japan............ | 1,000 h. p. | " | " | 110 ft. head |
| Chollar Mining Company, Nevada....... ........... | 750 h. p. | " | " | 1,680 ft. head |
| San Antonio Electric Power Co., Cal ................ | 800 h. p. | " | " | 400 ft. head |
| Standard Con. Mining Co., Bodie, Cal................ | 650 h. p. | " | " | 340 ft. head |
| Coeur d'Alene Silver Mining Co., Idaho.............. | 760 h. p. | " | " | 810 ft. head |
| Belmont Con. Mining Co., Colorado................. | 250 h. p. | " | " | 610 ft. head |
| Mammoth Mine, Madera County, Cal................. | 175 h. p. | " | " | 60 ft. head |
| Glenwood Light and Power Co., Colorado.......... | 450 h. p. | " | " | 380 ft. head |
| Casapalca Electric Light Co., Casapalca, Peru... | 400 h. p. | " | " | 170 ft. head |
| Electric Lt. and Power Co., San José, Costa Rica | 400 h. p. | " | " | 200 ft. head |
| Mt. Lowe Railway Company, Altadena, Cal....... | 200 h. p. | " | " | 1,250 ft. head |
| Revenue Tunnel Co., Ouray, Colorado................ | 600 h. p. | " | " | 650 ft. head |
| South Yuba Canal Co., Newcastle, Cal................ | 130 h. p. | " | " | 420 ft. head |
| Central Cal. Electric Co., Newcastle, Cal........... | 1,200 h. p. | " | " | 420 ft. head |
| Roaring Fork Elec. L. & P. Co., Aspen, Col......... | 1,400 h. p. | " | " | 820 ft. head |
| Cia. de Luz Electrica, San Salvador, C. A.......... | 300 h. p. | " | " | 60 ft. head |
| Santa Ana Elec. Co., San Salvador, C. A. .......... | 400 h. p. | " | " | 76 ft. head |
| Medillin Elec. Lt. Co., Medillin, U. S. C............ | 700 h. p. | " | " | 340 ft. head |
| Cia. Esplotadora de Lota y Coronel, Chili, S. A... | 650 h. p. | " | " | 360 ft. head |
| F. D. Mendiola, Boza, Costa Rica, C. A.............. | 400 h. p. | " | " | 200 ft. head |
| Cartago Electric Light Co., Costa Rica, C. A...... | 300 h. p. | " | " | 250 ft. head |
| Moodies Mining Co. Limtd., South Africa.......... | 800 h. p. | " | " | 130 ft. head |
| Honolulu Elec. Light Co., Honolulu, H. I......... | 100 h. p. | " | " | 200 ft. head |
| Bozeman Elec. Light Company, Montana........... | 170 h. p. | " | " | 124 ft. head |
| Wallace Elec. Light Co., Wallace, Idaho............ | 125 h. p. | " | " | 124 ft. head |
| Bell Electric Light Company, Auburn, Cal....... .. | 100 h. p. | " | " | 80 ft. head |
| Alaska Gold Mining Co., Douglas Island, Alaska | 150 h. p. | " | " | 460 ft. head |
| Banner Mining Company. Oroville, Cal............ | 160 h. p. | " | " | 120 ft. head |
| Co-operative Mining & Milling Co., Arizona....... | 100 h. p. | " | " | 150 ft. head |
| Helena & Livingston Smelting Co., Montana...... | 600 h. p. | " | " | 725 ft. head |
| Burmah Ruby Mines, Mandalay, India............... | 400 h. p. | " | " | 120 ft. head |
| Fairhaven Electric Co., Fairhaven, Wash........... | 120 h. p. | " | " | 300 ft. head |
| Phoenix Mining Co., New Zealand, Australia...... | 200 h. p. | " | " | 180 ft. head |
| Bear Valley Electric Power Co., Nova Scotia...... | 270 h. p. | " | " | 170 ft. head |
| Tinebi El. Lt. & Power Co., Costa Rica, C. A...... | 300 h. p. | " | " | 260 ft. head |
| Weaverville El. Light & Power Co., Cal........... . | 700 h. p. | " | " | 200 ft. head |
| Mullan El. Lt. & Power Co., Mullan, Idaho......... | 160 h. p. | " | " | 170 ft. head |
| Calumet Mining Company, Shasta Co., Cal......... | 300 h. p. | " | " | 800 ft. head |
| Delmatia Mining Co., Eldorado Co., Cal............ | 230 h. p. | " | " | 110 ft. head |
| Gold King Mining Company, Colorado............... | 1,200 h. p. | " | " | 500 ft. head |
| Sheridan-Belmont Mining Co., Colorado............. | 360 h. p. | " | " | 240 ft. head |
| Barrio-Nueva Jute Company, Orizaba, Mexico..... | 700 h. p. | " | " | 100 ft. head |

# LIST OF ELECTRIC POWER INSTALLATIONS.

|  |  | | |  |
|---|---|---|---|---|
| Southern Cal. Power Co., Santa Ana Cañon, Cal. | 5,000 h. p., | running | under | 700 ft. head |
| Cia. de Papal de San Rafael y Anexas, Mexico | 1,200 h. p., | " | " | 950 ft. head |
| Cia. de Papal de San Rafael y Anexas, Mexico | 550 h. p., | " | " | 220 ft. head |
| Hiroshima Electric Light Co., Hiroshima, Japan | 1,200 h. p., | " | " | 240 ft. head |
| Utah Power Company, Salt Lake City, Utah | 2,000 h. p., | " | " | 440 ft. head |
| Ned. Ind. Mijnbouw Mtiji., Celebes, East Indies | 650 h. p., | " | " | 550 ft. head |
| Petropolis Electric Light Co., Brazil, S. A. | 1,000 h. p., | " | " | 200 ft. head |
| Cie. de Boa Vista, Diamantina, Brazil, S. A. | 400 h. p., | " | " | 350 ft. head |
| Cripple Creek District Ry., Cripple Creek, Colo | 400 h. p., | " | " | 550 ft. head |
| Burmah Ruby Mines, Mandalay, India | 220 h. p., | " | " | 60 ft. head |
| Miller Manual Labor School, Albemarle, Va | 70 h. p., | " | " | 225 ft. head |
| Diamond Hill Gold Mining Co., Townsend, Montana | 700 h. p., | " | " | 170 ft. head |
| Duplantier Elec. Light Co., San Jose, Costa Rica, C. A. | 150 h. p., | " | " | 52 ft. head |
| San Ildefonso Paper Mill, San Ildefonso, Mexico | 1,000 h. p., | " | " | 185 ft. head |
| Yuba Electric Power Company, Marysville, Cal. | 2,000 h. p., | " | " | 290 ft. head |
| Crested Butte Light & Water Company, Colo | 75 h. p., | " | " | 300 ft. head |
| Development Syndicate, Oroville, California | 83 h. p., | " | " | 120 ft. head |
| Kyoto City—Electrical Department—Kyoto, Japan | 150 h. p., | " | " | 100 ft. head |
| Alumbrado Elec. de Quezaltenango, Guatemala, C. A. | 140 h. p., | " | " | 83 ft. head |
| South Yuba Water Company, Newcastle, Cal. | 134 h. p., | " | " | 440 ft. head |
| Concheno Mining Company, Concheno, Mexico | 260 h. p., | " | " | 190 ft. head |
| Santa Ysabel Mining Company, Jamestown, Cal. | 80 h. p., | " | " | 130 ft. head |
| Nevada County Electric Power Co., Nevada City, Cal. | 1,600 h. p., | " | " | 200 ft. head |
| Juneau Electric Light Company, Juneau, Alaska | 200 h. p., | " | " | 225 ft. head |
| Hilo Electric Light & Power Company, Hawaii, H. I. | 260 h. p., | " | " | 250 ft. head |
| Tuolumne Electric Light & Power Co., Jamestown, Cal. | 500 h. p., | " | " | 995 ft. head |
| Oroville Gas & Electric Company, Oroville, Cal. | 75 h. p., | " | " | 100 ft. head |
| Cooperative Mining & Milling Co., Bumblebee, A. T. | 75 h. p., | " | " | 150 ft. head |
| Empress Electricio de la Antigua, Guatemala, C. A. | 200 h. p., | " | " | 65 ft. head |
| Spring Creek Electric Power Company, Shasta Co., Cal. | 300 h. p., | " | " | 800 ft. head |
| Telluride Power Transmission Co., Telluride, Colo | 900 h. p., | " | " | 901 ft. head |
| Cañon Creek Electric Company, Gem, Idaho | 50 h. p., | " | " | 90 ft. head |
| Columbia & Western Railway Company, Trail, B. C. | 550 h. p., | " | " | 267 ft. head |
| Sandon Water Works & Light Company, Sandon, B. C. | 200 h. p., | " | " | 400 ft. head |
| Fort Wayne Electric Corporation, Arizona | 60 h. p., | " | " | 200 ft. head |
| Waianæ Electric Company, Hawaiian Islands | 270 h. p., | " | " | 690 ft. head |
| Boca Ice Company, Prosser, California | 40 h. p., | " | " | 20 ft. head |
| Payson Electric Light & Power Co., Payson, Utah | 150 h. p., | " | " | 125 ft. head |
| Gold Hill Water Company, Virginia City, Nev. | 30 h. p., | " | " | 230 ft. head |
| Big Dipper Mining Company, Iowa Hill, Cal. | 20 h. p., | " | " | 230 ft. head |
| Gold Bluff Mining Company, Downieville, Cal. | 125 h. p., | " | " | 270 ft. head |
| Pioneer Mining Company, Plymouth, Cal. | 25 h. p., | " | " | 560 ft. head |
| Gold Dredging Company, Bannock, Montana | 150 h. p., | " | " | 350 ft. head |
| Caroline Mining Company, Ouray, Colo | 520 h. p., | " | " | 650 ft. head |
| Ouray Electric Light Company, Ouray, Colo | 350 h. p., | " | " | 250 ft. head |
| Hidden Treasure Gold Mining Co., Placer Co., Cal. | 200 h. p., | " | " | 810 ft. head |
| Jumper Mining Company, Stent, California | 400 h. p., | " | " | 230 ft. head |
| Mountain Copper Company, Keswick, Cal. | 400 h. p., | " | " | 240 ft. head |
| Ontario Silver Mining Company, Park City, Utah | 160 h. p., | " | " | 120 ft. head |
| Antigua Electric Light Company, Guatemala, C. A. | 280 h. p., | " | " | 65 ft. head |
| Silver Lake Mines, Silverton, Colorado | 300 h. p., | " | " | 180 ft. head |
| Santa Fe Water & Investment Co., Santa Fe, N. M. | 120 h. p., | " | " | 160 ft. head |
| Santa Gertrudis Mining Co., Orizaba, Mexico | 250 h. p., | " | " | 100 ft. head |
| Empresa Electrica Antigua, Guatemala, C. A. | 260 h. h., | " | " | 65 ft. head |
| Los Compania Electrica, Medillin, U. S. Colombia | 600 h. p., | " | " | 490 ft. head |
| Cia. Electrica San Cristobal, Venezuela, S. A. | 100 h. p., | " | " | 150 ft. head |
| Cia. de Luz Electrica de Heredia, Costa Rica, C. A. | 400 h. p., | " | " | 200 ft. head |
| San Jose Electric Light Co., Costa Rica, C. A. | 400 h. p., | " | " | 200 ft. head |
| Buttermilk Falls Electric Co., Ft. Montgomery, N. Y. | 200 h. p., | " | " | 85 ft. head |
| Ophir Mining Company, Ophir Hill, Utah | 100 h. p., | " | " | 108 ft. head |
| Alumbrado Electric Company, San Salvador City, C. A. | 750 h. p., | " | " | 100 ft. head |
| Neihart Water Company, Neihart, Montana | 75 h. p., | " | " | 310 ft. head |
| Tjimpaka Tea Estate, Island of Java, D. E. I. | 70 h. p., | " | " | 60 ft. head |
| Dutch East Indian Electric Light Company | 800 h. p., | " | " | 560 ft. head |
| Bear Valley Electric Company, Nova Scotia | 110 h. p., | " | " | 90 ft. head |
| Santa Ana Electric Company, San Salvador, C. A. | 200 h. p., | " | " | 60 ft. head |
| Talemanco Electric Light Company, Venezuela, S. A. | 160 h. p, | " | " | 200 ft. head |
| Mendoza Electric Light Company, U. S. Colombia | 110 h. p., | " | " | 76 ft. head |
| Bridgetown Electric Light Co., N. S. W., Australia | 140 h. p., | " | " | 126 ft. head |
| Rossland Electric Light Company, Rossland, B. C. | 250 h. p., | " | " | 240 ft. head |
| British Columbia Electric Railway. Victoria, B. C. | 1,200 h. p., | " | " | 570 ft. head |

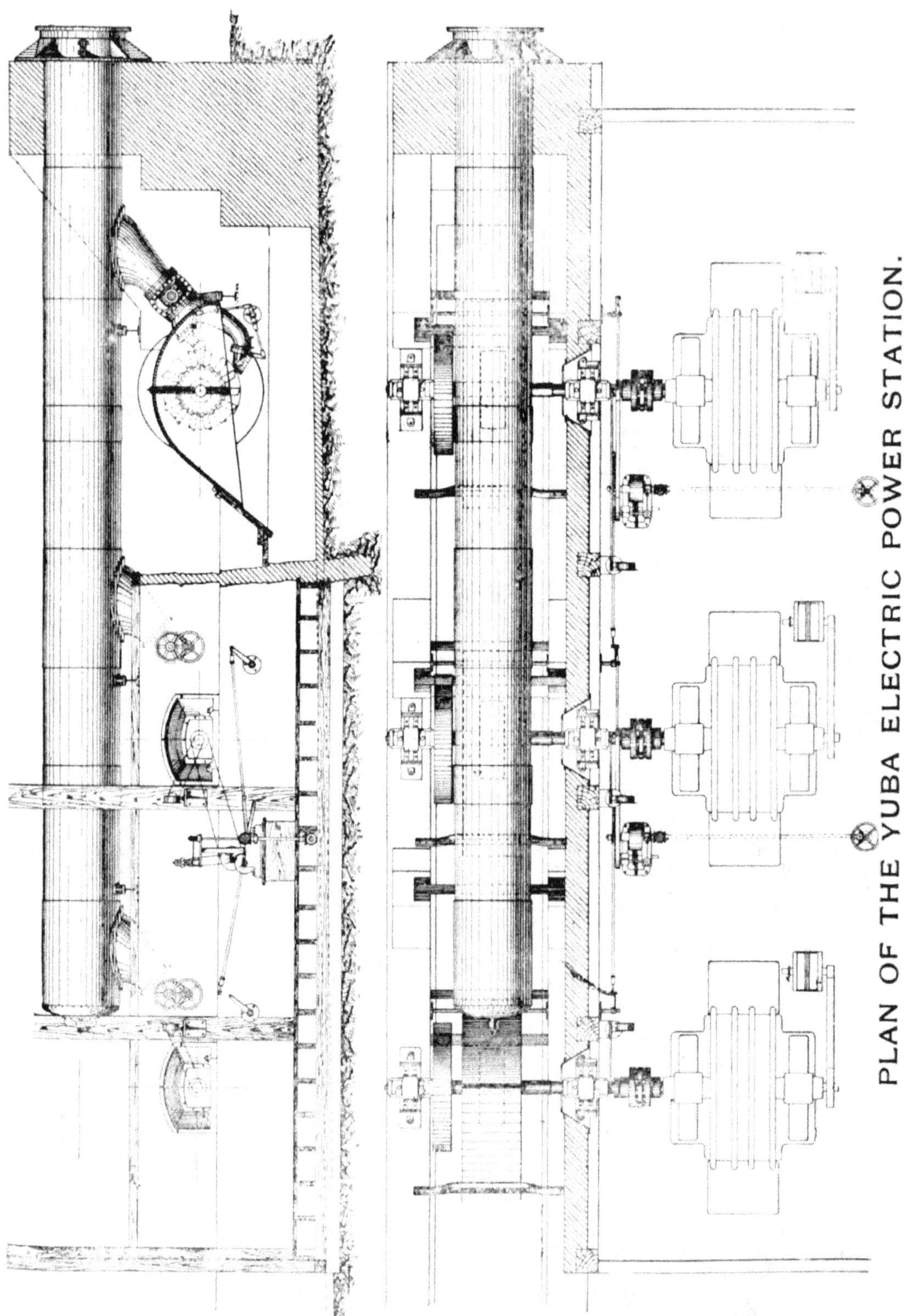

PLAN OF THE YUBA ELECTRIC POWER STATION.

The above plan shows the arrangement of a station where the wheels are located outside the main building, directly under the receiver. The shafts run through the wall and connect with the generators by flexible insulated couplings. For a large station, this is an excellent arrangement, involving less expense than iron-clad wheels with bed-plates, located in same room with generators.

## ELECTRIC POSSIBILITIES AND LIMITATIONS.

The question of electric power transmission must always be determined by the conditions of each particular case. Every proposition of this character is an engineering problem in itself, to be carefully considered and worked out after a full investigation of all the facts and circumstances connected with it. In a general way it may be said that long-distance transmission of power is as much an assured fact as the long-distance telephone, with which all are now so familiar.

There is, however, much difference of opinion among electrical authorities as to the distance electric power can be safely and economically transmitted, some contending that the Niagara Falls plant can supply power, within a radius of 200 miles, cheaper than it can be produced at any point within that distance by steam-engines of the most economical type, with coal at $3.00 per ton.

This, it may be said, is generally regarded as an extravagant estimate, and not supported by the more conservative of the profession, though the possibility of transmission even a much greater distance in the near future is conceded.

Success in the past has followed so closely upon prediction, that nothing in this way now promised is held to be altogether improbable.

### STREAMS THAT ADMIT OF UTILIZATION

or power purposes, with much less expense and under more favorable conditions than at Niagara, are so numerous in all parts of the country that there is no occasion —for some time at least—to consider a proposition involving such an extreme distance.

Numerous localities may be referred to, especially in the mountainous or mining regions of this and many foreign countries, where power is vastly more valuable than in the older and more settled communities, and where higher water heads are available, which admit of development at a much smaller cost than with turbines under low heads.

The distance to which power can be transmitted and made a commercial success depends upon a variety of conditions, such as the cost of producing the power, its value at point of delivery, etc., etc. The loss in electric transmission of a given power is determined by the voltage carried and size of wire, varying under ordinary circumstances from 5 to 20 per cent. For a short distance, the more common practise is a direct current with low voltage; for a longer distance, the multiphase alternating system, with a high voltage.

Long-distance transmission is simply a question of potential and insulation. With improvements recently made in insulators, a pressure of 25,000 to 30,000 volts is now considered entirely practicable. This will cover a distance of 75 to 100 miles with reasonable economy as to loss of current and cost of wire.

### SOME THINGS IN THIS CONNECTION

may be considered as settled facts.

The transmission of electric power is in no sense an experiment, and is entirely practicable in any case within any reasonable limit of distance, without regard to local conditions.

Electric power is fully as reliable as steam, while under better control and much more easily distributed and applied to various mechanical uses. This is especially true in its application to all mining operations.

The running expenses and cost of maintenance of a plant of this character are but a fraction of those of a steam plant, while the percentage of depreciation is very much in the same ratio.

PELTON WHEELS are now running with but few exceptions every electric station in this as well as most foreign countries, where conditions favor as to head, and in every instance they have been a most pronounced success, making profitable many enterprises that could not be run by steam power without loss.

The statement is made, with a full knowledge of the facts, that there is not a station in this country of any importance operating other tangential wheels, that is running with any degree of success or satisfaction. In fact, several may be mentioned that are conspicuous failures, both as to efficiency and regulation.

INTERIOR OF THE BIG COTTONWOOD POWER STATION.

## BIG COTTONWOOD POWER COMPANY.

One of the most important Electric Power installations in the country is now in successful operation in Utah. The station referred to is located in the big Cottonwood cañon, 14 miles from Salt Lake City, the stream from which affords a supply of 4,000 cubic feet of water per minute at the lowest stage, and which, under the head available, develops 2,500 horse-power. The water is brought to the wheels through 2,400 feet of 50-inch pipe, made of plate steel varying from ¼ to ¾ inch in thickness, made in sections of 32 feet, which are connected by steel flanges.

The power station consists of four 60-inch PELTON WHEELS—capacity of 650 horse-power each—running under 370 foot head at 300 revolutions. These wheels are direct connected to four General Electric Generators, three-phase type, both mounted on same bed-plate. Special wheels are provided for running the exciters, which are also direct connected.

The power thus generated is transmitted to Salt Lake City and used for lighting and general power purposes. An additional water supply is available, which it is proposed to utilize later, increasing the capacity of the station to 5,000 horse-power.

These works have involved an outlay of something over $500,000, which has been furnished mostly by local capitalists. The company is in receipt of a large permanent income from sale of power sufficient to provide for its bonded indebtedness and insure remunerative dividends to stockholders. The plant as a whole is regarded as one of the most successful of the kind yet made, and has realized in its operation the most sanguine expectations of its projectors.

The financial as well as engineering success that has attended this as well as most enterprises of similar character in various parts of the country, has given assurance of safe and profitable returns on all such investments when made with reference to proper hydraulic and electric appliances for securing the best results, cost of works and commercial value of power produced.

The illustration on the opposite page affords a good idea as to the interior arrangements of this plant.

## THE UTAH POWER COMPANY.

This plant is located on the Big Cottonwood Creek, some three miles below that of the Big Cottonwood Company, the water supply being taken from the tail-race of that company. The station originally consisted of a special 59-inch double-nozzle PELTON WHEEL, developing 1,200 horse-power, direct connected to a Westinghouse generator of the same capacity—both being mounted on a heavy cast-iron bed plate. It was found later that in order to meet the demands for power, it would be necessary to increase the capacity, and a complete duplicate of both wheel and dynamo was added, thus giving a total capacity in the plant of 2,400 horse-power.

The wheels run under a head of 440 feet at 330 revolutions and weigh 6,000 pounds each, they being made extra heavy for fly wheel effect to facilitate regulation. Special wheels run the exciters, they being also direct connected. Housings of wheels are all made of heavy steel plate and fitted with stuffing boxes to prevent leakage.

Power thus generated is transmitted to Salt Lake City, some twelve miles distant, and used for light and general power purposes, the entire system of electric railways in Salt Lake City being operated from this circuit. The variations in load are necessarily sudden and heavy, but the governing apparatus affords close and sensitive regulation. The plant is in every way a success and its operation is wholly satisfactory to all concerned.

## YUBA ELECTRIC POWER COMPANY.

This station is located near Dry Creek, twenty miles northeast of Marysville, Cal. The water is first conveyed through 9 miles of flume and 23 miles of ditch, then carried to the power-house through 850 feet of 42-inch riveted sheet steel pipe, affording an effective head of 290 feet. The pipe line terminates in a receiver 42 inches in diameter, to which are attached three branches to wheel connections.

Three double-nozzle PELTON WHEELS 40 inches in diameter, running at 400 revolutions, are direct connected by flexible couplings to the same number of Stanley two-phase generators, each of 500 horse-power capacity.

The power is transmitted by pole line to Marysville, a distance of 20 miles, and used for lighting the city and general power purposes. There is also an 8-mile circuit delivering lights and power for mining purposes at Brown's Valley—both power and lights being taken from each circuit.

The Stanley dynamos generate 2,400 volts, and the voltage is then transformed up to 16,500, which is transmitted on the line, then reduced to 2,400 volts at the sub-stations. The line is equipped with petticoat glass insulators on main circuit throughout, mounted on iron pins with porcelain bases.

The Pelton wheels are controlled by electric relay governors, which act upon deflecting hoods fitted to the nozzles, each wheel shaft having a fly-wheel of 6,000-lbs. weight to facilitate regulation.

This plant has some new and many interesting features, both in generation and distribution of power, and its successful operation is eliciting much comment among parties interested in projects of this character.

## ELECTRIC POWER TRANSMISSION.

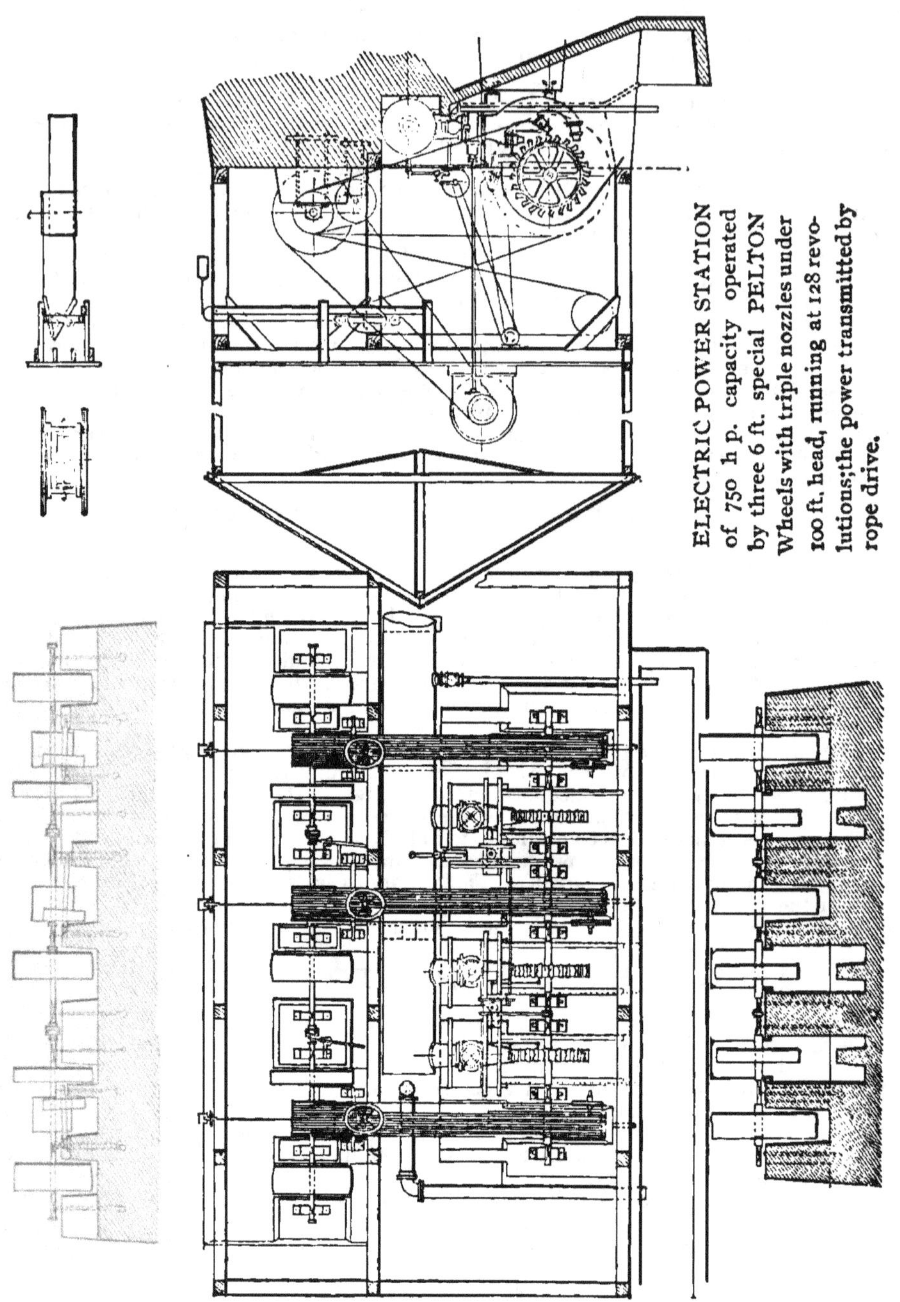


## PLAN OF ROPE DRIVE POWER STATION.

The above plan shows a station operated by a rope drive—the head being too low to admit of a direct connection to generators. The wheels are set on concrete or masonry foundations, and are located in a separate compartment, the rope drive running through the wall to sheaves on generator shafts.

## ELECTRIC POWER TRANSMISSION.

The Roaring Fork Electric Light and Power Company's plant at Aspen, Colorado, affords a most interesting example of the application of water-power to the production and distribution of electrical energy, and the convenient and profitable use made of it in mining operations. This was one of the first attempts on a scale of any magnitude to operate the various machinery required in mills and mines by electric transmission, and the success that has attended the experiment has attracted wide attention in all parts of the country, and has been the means of its adoption for similar service in many other localities, with equally satisfactory results.

"The power station is located near the mining town of Aspen, in the heart of the Rocky Mountain range—a place of some 7,000 inhabitants—having an elevation of nearly 8,000 feet. Water is brought to the station in a single line of pipe, consisting of 500 feet of 16 inch, and 3,500 feet of 14 inch—which discharges into a receiver, from which short connections are made to the wheels. The power plant consists of eight 24-inch PELTON WHEELS, which run at 1,000 revolutions under a head of 820 feet with a maximum capacity of 175 horse-power each, aggregating some 1,400 horse-power.

"Each wheel runs a separate generator, and connection is made by belt direct, the power developed being made to conform to the requirements by the use of reducing tips, so that only as much water is applied to the wheels as is necessary to run the generators at proper speed.

"The LIGHT plant is run by three Brush arc dynamos of capacity of 60 lights each (2,000 c. p.), two Brush incandescent dynamos of 450 lights each (16 c. p.), three Westinghouse alternating current machines of 750 lights each (16 c. p.). The entire town of Aspen is lighted from this station, as well as many mills, mines, and sampling works in the vicinity.

"The POWER plant consists of one Edison power generator, 80,000 watts, 500 v., 160 amp., and one Edison power generator, 40,000 watts, 500 v., 80 amp. The power thus generated is used for operating mills, hoist, pumps, tramways, etc., as well as various other mining operations within a radius of 3 to 4 miles from the generating station.

"This station has been in continuous operation for nine years, with practically no expense in the way of repairs nor interruption of any moment to the service, which, considering the severity of the weather encountered in such an altitude a part of the year, gives the stamp of reliability and efficiency to the PELTON SYSTEM OF POWER, as well as to its transmission by the various electrical means described."

NOTE.—The wheels running the above plant weigh but 90 lbs. each, and therefore show a capacity of nearly 2 h. p. for every pound of material, and including accessories to make the plant complete, such as shafting, boxes, pulleys, gates, nozzles, etc., the proportion would be 4½ lbs. to every h. p. developed. The relative proportion of material in the best type of steam plants is from 400 to 500 lbs. for every h. p.

### NEW WORKS OF THE ROARING FORK COMPANY.

This station is located so as to utilize the waters of both Maroon and Castle Creeks, the water being brought from the former in a 26-inch pipe 2,500 feet in length, and delivered to the wheels under a head of 312 feet, and from the latter in a 24-inch pipe 4,300 feet long, and delivered to the wheels under a head of 330 feet.

The power plant consists of five PELTON WHEELS of 250 h. p. capacity each. The wheels are about five feet in diameter, being varied somewhat in size to give the same speed—300 revolutions—under the different heads the pipe lines afford.

The wheels are all on one shaft, and provided with friction clutch couplings, making them operate independently. The electric installation consists of five 250 h. p. Brush generators which are driven direct from the main line shaft by friction clutch pulleys.

This plant is operated as a general distributing station, supplying power to various mines within a distance of about five miles, and its operation has been in every respect as satisfactory as the original plant of this company above described.

### PEOPLE'S ELECTRIC LIGHT AND POWER COMPANY.

This station is located on Castle Creek, about one mile from the town of Aspen, Colorado, and consists of two 5-ft. double nozzle PELTON WHEELS of capacity of 300 h. p. each, running under 180-ft. head at 240 revs. Also two 3-ft. double nozzle PELTON WHEELS of 75 h. p. capacity each, running under the same head at 345 revs.

The electrical equipment was furnished by the General Electric Co., and consists of four 100 K. W. Edison generators, which are driven direct from the wheel shafts, also one 50 K. W. T. & H. alternating dynamos and one 25 K. W. arc dynamo. The four large machines are for power purposes, and the two small for light. Doolittle governors, acting upon balance valves in the nozzles of the wheels, afford perfect regulation. Power is furnished from this station to various mills and mines within a radius of about three miles.

The wheels and connections, as well as the transmitting machinery, were erected and delivered in complete running order by the PELTON WATER WHEEL CO. This plant embraces the latest and most improved machinery—both hydraulic and electric—and is regarded as one of the most noted successes yet made in the application of electricity to mining work.

DAM OF THE NEVADA COUNTY ELECTRIC POWER COMPANY.

The above print shows the means ordinarily adopted to divert the water of a stream into a flume or canal, with which the pipe line is connected, running to the power station. A head gate is located at the inlet for shutting off the water when required, with a waste way as shown.

The pipe line may connect direct with the head gate when conditions favor such an arrangement. In many cases, a flume or canal may be carried along the bank of a stream a considerable distance with but small loss of head, reducing in this way the length and consequent cost of the pipe line.

## NEVADA COUNTY ELECTRIC POWER COMPANY.

This station is located on the South Yuba River, 6 miles from Nevada City, and consists of two sets of Pelton wheels, 750 h. p. each, which run under a head of 206 feet at 400 revs. The wheels are direct connected to Stanley dynamos of same capacity, which generate a current of 5,500 volts. The power thus furnished is transmitted over a line—the extreme length of which is 12½ miles, the towns of Nevada City and Grass Valley being furnished with light, and several mills in the vicinity with both light and power. Satisfactory regulation is afforded by electric relay governors.

To admit of direct connection under so low a head, three wheels were required for each unit of power. All are on one shaft and in one housing, as shown in print below. Water is brought from the Yuba River through 4 miles of flume and thence to the station by 350 feet of pipe. The water operating this plant is exceptionally bad, carrying at times from 10 to 20 per cent of slimes from mill tailings, very destructive to wheels. By a special construction, the wear from this cause has been reduced to a minimum.

## WANOOSNOCK ELECTRIC POWER COMPANY.

This station is located about 1½ miles from the city of Fitchburg, Mass. Water is obtained from a stream which affords a minimum supply of some 2,000 cubic feet per minute. A pipe line 1,500 feet in length laid along the bank of the stream to the power station, gives a fall of 200 feet. The plant consists of six double-nozzle PELTON WHEELS 33 inches diameter keyed to one shaft, speeded at 375 revolutions, and developing 100 horse-power each.

Two 300 horse-power Westinghouse generators are connected direct to the wheel shaft, one at each end. The wheels are inclosed in iron cases and mounted on steel beams, upon which the generators are also placed, affording a secure foundation and accurate alignment to the entire plant. The wheels were made of this small diameter in order to give the generators the necessary speed by direct connection, otherwise the entire power could have been obtained from a single wheel.

A part of the power furnished by this station is used by the Simmons Saw Works, and the remainder for general light and power purposes.

MANITOU TUNNEL POWER PLANT, COLORADO.

The above print shows a PELTON WHEEL direct connected to the armature of a dynamo with a flexible insulated coupling, both set on a concrete foundation. The most approved practice, however, is to mount both wheel and dynamo on a channel iron frame or cast-iron bed plate. This affords a perfect alignment without absolute dependence upon the integrity of the foundation.

## SAN JOAQUIN ELECTRIC POWER COMPANY.

One of the most successful as well as remarkable electric power installations yet made in any part of the world is that of the above-named company, located some 35 miles from the city of Fresno, Cal.

The water supply is taken from the North Fork of the San Joaquin River, at a point where it flows down a rocky canyon filled with rapids and cataracts, with high, precipitous mountains on either side.

At the head of these rapids, the water is carried in a canal along the summit of a high ridge, a distance of 6 miles, to a point some 1,500 feet above the river. A reservoir has been constructed on the summit of the mountain, covering 8 acres, to a depth of 10 feet, which holds sufficient water to run the entire plant 6 or 7 days in case of accident to the canal which would temporarily shut off the supply.

A pipe line 4,000 feet long, connecting with the canal on the summit, runs to the foot of the mountain, where the power station is located. The pipe is 24 inches diameter at the upper end, 18 inches at the lower, and is made of plate steel ¼ inch to ⅝ inch in thickness, graduated to the pressures it sustains at different points on the line.

The power station consists of three 500 horse-power PELTON WHEELS, 57 inches diameter, which run under a vertical head of 1,410 feet. Each of these wheels is direct connected to the same number of 300 K. W. generators, speeded at 600 revolutions—General Electric three-phase type. Two 20-inch PELTON WHEELS of 20 horse-power capacity are also provided for running the exciters. Fly wheels 6 feet in diameter, weighing 6,000 pounds, are attached to the wheel shafts.

Perfect and absolute regulation is afforded by Relay Governors, one of which is attached to each wheel, giving separate and independent control, and providing for all changes in load due to operating a variety of machinery in an irregular way.

The power thus generated is transmitted to the city of Fresno, 35 miles distant, over 6 ⅜-inch copper wires, at 19,000 volts, furnishing power for the railroad, electric light, and pumping plants of the city; also for various manufacturing purposes.

From this point, the line has been extended to Hanford, 32 miles distant, making the entire transmission 67 miles—with one exception the longest distance of any line thus far erected.

This station has now been running upwards of two years with a high degree of efficiency and economy, practically without repairs and without any interruption to continuous service.

The features of special interest connected with this installation are, the large amount of power developed, distance of transmission, the variety of purposes to which the power is applied, and the enormous head under which the wheels are running.

The operation of this plant—in many respects without precedent—has been so reliable, economical, and successful as to inspire the utmost confidence in the practicability of long-distance electric power transmission, even under exceptional and untried conditions.

## SPRING CREEK ELECTRIC POWER COMPANY.

This station is located near the town of Copley, Cal., and consists of a 29-inch PELTON WHEEL direct connected to a General Electric generator.

The wheel runs under a head of 800 feet, and develops 300 horse-power. The water is brought from Spring Creek through 3,350 feet of riveted steel pipe, varying from 10 to 16 inches in diameter.

The power thus furnished is transmitted from 2 to 5 miles, and is used to run mills, hoist and other mining machinery, as also a mine railroad.

The plant runs continuously without any skilled attention and with practically no expense in maintenance, and affords an excellent example of the advantageous use of water-power in connection with all mining operations.

## MANITOU TUNNEL POWER PLANT.

This station, an illustration of which will be found on the opposite page, is located a short distance above the Iron Springs Hotel, at Manitou, Colorado, and consists of a 500 horse-power PELTON WHEEL, direct connected to a General Electric Company's generator. The wheel runs under a head of 600 feet, and is speeded at 600 revolutions.

The power thus generated is carried a distance of 8 miles, and runs an air-compressor for supplying the drills, operating in what is known as the Strickler Tunnel, which is being driven through a spur of Pike's Peak range. This tunnel is 6,400 feet long, and forms a part of the new water-works system of Colorado Springs. Operations on the tunnel are carried on from both ends, and light as well as power is supplied from the station for the power-house as well as all underground work.

This is believed to be the first instance in which the resources of water-power and electricity have been brought to bear upon a project of this character. The facility it has afforded for the rapid and economical prosecution of the work has been a gratifying surprise to all interested in the enterprise.

The 13-inch main, supplying Colorado Springs, carries a pressure at the reservoir of 170 lbs., which it is now proposed to utilize, by means of PELTON WHEELS and electric generators, for lighting the city.

ELECTRIC POWER PLANT, STANDARD MINING COMPANY, BODIE, CAL.

## POWER PLANT—STANDARD MINING CO.

The cut on the opposite page shows an application of PELTON WHEELS to a high-speed generator by direct connection to the armature shaft. The wheels—four in number—are inclosed in two separate housings, all mounted on one shaft and connected to a Westinghouse generator by an insulated coupling. They are 21 inches in diameter, and have a combined capacity of 400 horse-power, with a speed of 860 revolutions under a running head of 340 feet. The wheels in the plant here referred to were made of this small diameter to give proper speed to the generator under the head in this case available. A slower speed generator—in conformity to modern practice—or a higher head of water, would have admitted of larger wheels, and consequently a less number to give the same power.

The current thus generated is transmitted to the works of the company, 12½ miles distant, at a pressure of 11,000 volts. It is used for running a 60-stamp mill, pumps, hoists, &c. Fuel was exceptionally high in that locality, and it was necessary to obtain cheaper power to admit of carrying on their operations on any extensive scale.

The manager, in his report to the company, states that the cost of the entire plant was $38,000, and that the saving effected over the previous operation of the plant by steam averaged some $1,500 per month—thus returning the entire outlay in about two years.

The statement is further made that in the four years the plant has been in operation no trouble has been experienced from the line, though running over a rough and mountainous country, with winter storms of great severity. Also, that no repairs have been required on either wheels or pipe line, and that the service, including regulation, has been at all times thoroughly reliable and efficient.

Considering the fact that this installation was one of the very first made for mining purposes under new and untried conditions, its success has been the more remarkable.

## ELECTRIC POWER TRANSMISSION.

The following account of the Aspen Mining and Smelting Co.'s plant, located at Aspen, Colorado, is taken from a paper read by M. B. Holt, M. E., before the American Institute of Mining Engineers:—

"The power station is located at Aspen, Colorado, and has a capacity of 100 electrical horse-power, which is obtained from two 50-horse-power Thomson-Houston motor type dynamos wound for a constant potential of 500 volts.

"These are operated by two double nozzle PELTON WATER WHEELS, 42 inches in diameter, running under a head of 80 feet The fall is obtained by fluming the water a distance of 1,300 feet, it being carried from thence to the power-house in a sheet-iron pipe. The wheels are set in concrete pits, and housed by iron hoods, which can be taken off with little trouble, allowing free access to the wheels.

"The generating station is 6,000 feet from the entrance to the tunnel, and the stationary motors are located underground at distances of 1,000, 1,200 and 1,800 feet from the entrance. From the power station to the tunnel, the current is carried by bare 00 copper wire, except at a distance of about 300 feet at each end of the line, where an underwriter's insulated wire of same size is used. Inside the mine the current is carried to the two main hoisting stations at distances of 1,000 and 1,200 feet by kerite seven-strand conductors. Each of the two hoists are raising 250 tons up a 60-degree incline 250 feet long every 24 hours. On none of the circuits inside the mine is there a loss exceeding five per cent. On the outside circuit, when the wires are carrying their maximum load, the loss does not exceed from five to six per cent.

"The main working galleries of the mine are lighted by electricity, the current for which is taken from the power mains, and five 100-volt incandescent lamps are connected up in series, eight horse-power of electrical energy being now employed for lighting purposes.

"The recognized advantages of electrical power for mining operations may be briefly summarized as follows:—

"1. It can be transmitted long distances with small loss, thus making it possible to use power at such a distance from its source as would render it otherwise unavailable, as in the case before us.

"2. The conductors for conveying electrical power from one point to another require less space, are more easily put in place and repaired, are easily tapped for branch circuits, and form a more flexible system throughout than any other mode of transmission permits.

"3. The electrical system is ideal, viewed from the standpoint of cleanliness.

"4. The stations for electrical power can be made to occupy a minimum space.

"5 If this system does not assist ventilation, it does not, on the other hand, vitiate the air in the mine workings.

"After five years' use, under the varying conditions of mining work, the electric current of 500 volts has proved itself free from danger to life, and has caused no inconvenience. The best illustration of the convenience and flexibility of the system is the diamond drill, where the conductors are unwound and strung up as the drill moves along, or taken down and coiled up, as may be desired."

AN ENGINEER OF THE PELTON WATER WHEEL COMPANY AND HIS CORPS OF ASSISTANTS MAKING A SURVEY FOR AN ELECTRIC POWER STATION IN THE MALAY PENINSULA.

## PLANT OF COMPANIA DE BOA VISTA BRAZIL.

This company owns some famous diamond mines near Diamantina, Brazil. Having no water at the mines, they could not be worked in the ordinary way; recourse was, therefore, had to a pumping plant. Some five miles distant a water-power was available, affording a head of 350 feet, requiring 3,000 feet of 20-inch pipe. The station consists of a special 6-foot PELTON WHEEL, which develops 450 h. p. and drives two General Electric 150 K. W. generators. The current from these machines is transmitted five miles to where the pumps are located, operating motors which are connected by means of gearing to two pair of Worthington-Duplex high-pressure pumps.

This station is a success and affords means of carrying on the mining operations of the company, which could not be made available in any other way.

### COMPANIA DE PAPAL DE SAN RAFAEL Y ANEXAS, MEXICO.

This plant is located at Amecameca, near the City of Mexico, and comprises two installations. One, consisting of two special PELTON WHEELS 11 feet 2 inches in diameter operating under an effective pressure of 420 pounds and running at a speed of 212 r. p. m. Each of these wheels has a capacity of 550 h. p. and is direct connected to 400 K. W. Westinghouse generators by flexible leather link couplings.

The power is transmitted a distance of about one mile to the paper mill of this company, where it is used for operating pulp grinders, rag beaters, Fourdrinier machines, etc. The pipe is 22 inches in diameter, 3,000 feet long, made of riveted steel ⅜ inch in thickness—the joints being connected by wrought steel flanges.

The second installation consists of two special 60-inch PELTON WHEELS working under a head of 200 feet at a speed of 212 r. p. m. These two wheels are mounted on one shaft inclosed in one housing, and connected by means of flexible leather link coupling to a 400 K. W. Westinghouse generator. The power from this plant is also transmitted to the company's mill.

### FABRICA DE SAN ILDEFONSO POWER STATION.

This plant is located near Tlalnepantla, Mexico, and was installed for the purpose of furnishing the woolen factory of San Ildefonso with light and power, the turbines formerly used being entirely inadequate. A canal 6,400 feet long carries 3,150 cubic feet of water per minute. From the canal a steel pipe line 700 feet long of 44, 46 and 48-inch diameter, runs to the power-house, delivering the water to the receiver, from which it is distributed through taper connections to the wheels under a head of 186 feet.

The power plant consists of three double-nozzle, 42-inch PELTON WHEELS, aggregating 1,000 horse-power. Each water wheel is direct connected to a 240 K. W. Westinghouse generator, developing a current at an initial pressure of 3,500 volts. At this voltage, it is transmitted to the Fabrica, 3¼ miles distant, where it is distributed to motors, each belted to its line-shaft for operating the mill. The various mill buildings, offices, etc., are also furnished with light from the same source.

This is regarded as a model plant, affording a reliable and economical power in a location where any extended operations are out of the question with a steam plant on account of the high cost of fuel.

### BRITISH COLUMBIA SMELTING COMPANY.

The largest smelting plant in the Northwest is that of the British Columbia Smelting and Refining Company, located at Trail, B. C. Power is supplied by two 36-inch PELTON WHEELS of 400 horse-power capacity, running under a head of 260 feet.

The water supply is obtained from Trail, Rock and Stony Creeks, which is collected in a reservoir about a mile above Trail, and from this point conveyed to the smelter through a line of steel pipe. The wheels run General Electric generators by belt connection.

This plant replaces an engine formerly used, the expenditure for which was deemed a good investment in a locality where fuel is abundant and reasonably cheap.

### DIAMOND HILL MINE POWER STATION.

This plant is located near the mining town of Hassel, Montana, and consists of four double-nozzle Pelton wheels 40 inches in diameter, all mounted on one shaft direct connected to the armature of a 500 K. W. General Electric generator, running at 300 revolutions. The wheels are inclosed in two separate housings, which—with bearings for the shaft and wheel nozzles—are mounted on a heavy iron I beam frame, which also supports the generator, bringing all in the same line.

The four double-nozzles are all connected by means of Y branches to the main pipe line, which is 42 inches in diameter and 900 feet long, delivering water to the wheel under a head of 160 feet. The transmission to the power-house is 1¾ miles, the line running over a very high mountain.

Excellent regulation is afforded by a relay governor operating on cut-offs attached to wheel nozzles, the wheel shaft carrying a 6,000-lb. fly-wheel.

The machinery driven at the mine consists of a 120-stamp mill, two large rock-breakers, 96 concentrators, a 75 horse-power compressor, and 500 lights about the mill and mine.

The plant as a whole is an unqualified success, and gives entire satisfaction to all concerned.

## AN IMPORTANT ENTERPRISE IN MEXICO.

The most interesting and important electric power installation so far made in Mexico is that of the Cia. Anonima de Transmission Electrica de Potencia, located in the state of Hidalgo, some one hundred miles north of the City of Mexico.

The water is taken from the Arroyo de Regla, a mountain stream having a minimum supply of 1,500 cubic feet per minute. A natural rock dam, at a favorable point in the cañon, impounds the water sufficiently to admit of its being diverted by a cut through the bluff into a canal, which follows mostly the contour of the mountain, a distance of 1½ miles, a work which involved the cutting of seven tunnels, aggregating a total length of 1,200 feet, through solid rock.

From the terminus of the canal, the water is carried to the power station through 1,700 feet of 30-inch pipe, which affords a vertical head of 810 feet, this being of varying thickness to correspond to the pressure at different points on the line, the lower portion being made of steel ¾ of an inch thick. The pipe line discharges into a receiver, 40 inches in diameter by 75 feet long, with which the wheels are connected by lateral branches. This is made of flange steel plates, ¾ of an inch thick, tested to 500 pounds water pressure, and weighs upward of 50,000 pounds.

The power station consists of five PELTON WHEELS, 40 inches in diameter, of a capacity of 600 horse-power each, direct connected to the same number of 12-pole, 3-phase generators, running at a speed of 600 revolutions, delivering the current at a pressure of 700 volts; also two 20-inch PELTON WHEELS speeded at 1,300 revolutions for running the exciters. The step-up transformers are wound for a ratio of 1 to 15, making the line potential a little over 10,000 volts at the generator end. There are three transformer sub-stations in which air-blast transformers are used. PELTON GOVERNORS attached to each wheel afford excellent regulation.

This station supplies power to the mines of the Real del Monte Company, one of the most extensive mining organizations in the world, employing upwards of 8,000 men. The power is used for operating mining machinery, such as stamp mills, crushers, pumps, hoists, ventilators, etc., etc. The mines of this company, said to be the richest in Mexico, are located within a radius of 20 miles, the maximum distance of power transmission being 23 miles, and the mean distance about 18 miles.

Various other mines in the vicinity are also supplied with power from this station, and the city of Pachuca furnished with light. A market for the entire power of the plant being thus afforded at highly remunerative rates, the financial success of the enterprise is assured; in fact, it is claimed that the entire outlay, some four hundred thousand dollars, will be returned to the company in two years' time, by the saving effected in fuel heretofore required in carrying on their various operations.

THE PELTON WATER WHEEL COMPANY, of San Francisco and New York, supplied the hydraulic part of the work; and the Foreign Department of the GENERAL ELECTRIC COMPANY, the electrical machinery.

This station has now been running some eighteen months, and with such success as to demonstrate beyond all question, the economy and advantages of this means of producing and distributing power, even where involving transmission a long distance over a rough and mountainous country, with application to a great diversity of service.

Considering the magnitude of the work, the extreme water pressure, the variety and extent of machinery operated, as well as the difficulties attending the transportation and erection of such massive machinery in a mountainous and almost inaccessible region, this may be regarded as one of the most remarkable electric power installations so far made in any part of the world.

## A BRAZILIAN ELECTRIC POWER PLANT.

This station is located about four miles from the city of Petropolis, Brazil, S. A., and consists of four PELTON WHEELS direct connected to General Electric Generators. The wheels run under a head of 262 feet and aggregate 1,200 horse-power. Separate wheels run exciters and all are controlled by PELTON GOVERNORS.

Water is brought to the station through 3,000 feet of 30-inch riveted steel pipe, which was shipped in sections and riveted up on the ground, forming a continuous line without joints.

The power thus developed is transmitted to Petropolis and used for lighting the city and general power purposes.

After running two years, a duplicate of the plant was ordered, bringing the capacity up to 2,400 h. p. The entire works are now in successful operation and are said to be extremely remunerative.

## NEDERLANDISCHE INDIANISCHE MIJNBOUW MTIJI.

This plant is located on the island of Celebes in the Dutch East Indies. The station consists of two special PELTON WHEELS, which run under 500 feet head at 300 revs., developing 800 h. p The wheels are direct connected to two high-tension alternating generators made in Holland. The power is transmitted some six miles and used for operating a gold mine, running stamp mills, pumps, hoists, an electric tramway, etc., etc.

A 31-inch PELTON WHEEL, direct connected to a 50 K. W. generator, speeded at 650 revs., is used for lighting purposes.

The pipe line is 18 inches in diameter and 12,000 feet long.

## AN EIGHTY-MILE POWER TRANSMISSION.

Altogether the most important electric power installation—as regards length of transmission—so far undertaken in any part of the world, is that of the Southern California Power Company, now under construction.

The station is located about eight miles from the town of Crafton, and consists of four 1,250 horse-power PELTON WHEELS, direct connected to the shafts of General Electric generators 750 K. W. each. The wheels are 82 inches in diameter, operate under 700 feet effective head, and run at a speed of 300 revolutions per minute. There are also three 24-inch PELTON WHEELS, capacity of 75 horse-power each, running at 1,000 revolutions per minute direct connected to exciters.

The water supply is taken from the Santa Ana River and Bear Creek—the point of diversion being at the junction of the two streams. It is thence conveyed through canal and flume some 17,000 feet—nearly one-half of this distance tunneled through solid rock. From the terminus of the canal to the power station, the water is carried in two 30-inch steel pipes, each 2,210 feet in length, which discharge into a receiver at the power-house 30 inches in diameter by 100 feet long. The receiver is made in two sections, connected in the center by a 30-inch cast steel gate. Each pipe line connects with a section of the receiver at right angles by lateral steel branches. The upper portions of the pipe lines are riveted sheet steel and the lower portions lap weld—the latter being one-half inch in thickness.

The transmission line is 80 miles long, composed of 2 circuits of No. 1 B. & S. gauge medium hard-drawn copper wire, supported on triple petticoat porcelain insulators, and is intended to carry a pressure of 33,000 volts. The power generated is to be transmitted to the city of Los Angeles, and used for running an extensive system of tramways, as also for lighting and general power purposes.

No doubt is entertained as to the entire success of this installation, which is estimated to cost about $900,000.

## TUOLUMNE COUNTY ELECTRIC PLANT.

This station is located near the town of Columbia, Cal., in the heart of a profitable mining section, and consists of a 500 horse-power PELTON WHEEL running under a head of 950 feet, direct connected to a General Electric Generator of same capacity.

The power thus developed is transmitted to various mines within a radius of five miles, running stamp mills, hoists, pumps, and all the various machinery connected with mining operations. The service has been at all times efficient, economical, and reliable.

## ELECTRICITY IN UNDERGROUND MINING

Electricity is now used in 300 mines in the United States, and proprietors of perhaps as many others are considering its utilization. Electricity tried under the arduous conditions of mining service has been shown to be peculiarly efficient, safe, and reliable. It presents a system of the greatest simplicity, completeness, and flexibility, permitting power from one source to be distributed in units of any desired size, and for any purpose, to the places where it can be employed to the greatest advantage, thereby securing the minimum consumption of power and expenditure of labor.

With electricity there is neither friction, heat, nor condensation. There is no leakage or loss of power when not in use, which especially recommends it for intermittent work. It is not affected by heat or cold, and does not vitiate the air. The rapid deterioration of timbering, a source of great expense in all mines, due to bad air or heat, is to a great extent obviated. It has been urged against electricity that it increased the risk of fire. As a matter of fact, it greatly decreases it, as the statistics of insurance companies testify.

## A FIFTY-THOUSAND VOLT TRANSMISSION.

(From the *Electrical Engineer*, New York.)

"It is not a fact generally known that a transmission at 50,000 voltage was some time ago tried for a period of two weeks on the line of the Telluride Power Station in Colorado, supplying the Gold King Mill, three miles distant. The first plant consisted of a single-phase 3,000-volt alternator, with direct transmission to a synchronous motor at the mill, since replaced by a three-phase transmission with step-up and step-down transformers. About the time the change was made the experiment was tried of transmitting at 50,000 volts.

"The transformers used were those now employed on the three-phase transmission there, said transformers being arranged to give a number of different voltages from 50,000 down according to the way they are connected.

"As stated, this transmission at 50,000 volts three-phase current was kept in service for about two weeks and no accidents occurred during the time. The line consisted of galvanized iron telegraph wires, supported on glass insulators. It was found that the self-induction afforded by the iron wire had a beneficial effect in counteracting the capacity of the line. The experiment was not continued for a longer time because the rainy season came on and proper provisions against lighting were not at hand. The transmission line is three miles long, and runs up a steep mountain side and over a very wild country."

## ELECTRIC PLANT ON THE COMSTOCK.

The Nevada Mill, formerly consisting of 40 stamps, with the usual complement of pans, settlers, etc., had heretofore been run with a 10-foot PELTON WHEEL from water supplied by the Gold Hill Water Company under a head of 460 feet. When the mill was enlarged to 60 stamps, with the additional amalgamating machinery—requiring in all some 400 horse-power—the question of securing this with a degree of economy that would admit of working the low-grade ores, was the problem. Fuel being scarce, steam power was out of the question. Water was also expensive, the only source of supply being brought from the Sierras through a pipe line 30 miles in length.

The Sutro Tunnel afforded an outlet some 1,700 feet below the surface, and the utilization of this immense pressure was decided upon in connection with electric transmission, as being the most feasible. A station was made by excavating a chamber at the Sutro Tunnel level, 1,680 feet below the surface, 25 feet in width, by 50 feet in length. In this were located six PELTON WHEELS 40 inches in diameter, connected to the shafts of the same number of primary generators, both running at a speed of 900 revolutions. The generator circuits all leading to the switch-board were so arranged that the current from any one could be thrown onto any of the outgoing circuits. From this point the wires are carried up the shaft to the motor room adjoining the mill, about one mile distant. The six electric motors to which the wires from the generators in the subterranean chamber are attached, are arranged in a single row parallel to the main driving shaft, with which they are connected in the usual manner. The surface PELTON WHEEL before referred to is also connected to the same shaft, the two distinct forces working together in perfect harmony.

This plant affords an interesting illustration of the double use of water and the saving effected thereby—something over four hundred horse-power being made available by this means from what may be termed waste water. It also affords the striking illustration of the power of water under extreme pressure. The wheels running without a load have a speed of 1,800 revolutions, and a peripheral velocity of 18,864 feet, or more than 3½ miles a minute. The water flows from the nozzle at the rate of 19,260 feet per minute, forming a stream nearly as solid as a bar of steel. The wheels develop 125 horse-power each, with a ⅝-inch nozzle, and have shown from 83 to 85 per cent efficiency. No repairs have been required other than occasionally replacing a bucket. This station has now been running upwards of seven years, and, though the first instalment of the kind on a scale of any magnitude, and made under most exceptional conditions, its operation has been in every way a success, fully meeting all expectations.

## THE BIG CREEK POWER COMPANY.

This station is located on Big Creek, in the Santa Cruz Mountains, Cal., and consists of two 500 horse-power wheels, direct connected to Westinghouse 2-phase generators, which run at a speed of 600 revolutions.

Separate wheels run the exciters, which are also direct connected. The water supply is first conveyed in a wooden flume 11,000 feet, and then brought to the station through 2,000 feet of 14-inch and 16-inch riveted steel pipe, affording a head at the power-house of 923 feet.

The power thus generated is transmitted over a No. five copper wire at 11,000 voltage to the city of Santa Cruz, 18 miles distant, where it is used for lighting, railway, and general power purposes.

No conditions offer so severe a test in regulation as running a railway in connection with a lighting system. The ELECTRIC RELAY GOVERNORS in this, as in all other instances where used, have proved equal to all demands made upon them, affording safe, sensitive, and satisfactory control.

### JUMPER MINE ELECTRIC POWER PLANT.

This station is located on Sullivan Creek, one mile from Stent, Tuolumne County, Cal., and consists of 4 feet double-nozzle PELTON WHEEL, mounted on a cast-iron bed-plate, set on masonry foundations. The wheel runs under a head of 230 feet, at a speed of 290 r. p. m., and drives, by belt connection, a 120 K. W. Westinghouse generator. The current is transmitted at 2,400 volts to the works of the company, 600 feet above the station and three-quarters of a mile distant, operating a 90 h. p. motor, which drives a Rix Air Compressor, furnishing power for a mine, hoist, drills, etc., etc.

All the underground workings of the mine, as well as offices and other buildings, are furnished with light. The water supply is brought through a line of 24-inch pipe, 2,900 feet in length.

This plant, though of moderate capacity and a short transmission, demonstrates the fact that a small amount of power can be produced and distributed by means indicated with a small outlay, and run at a nominal cost as compared with steam.

## CENTRAL CALIFORNIA ELECTRIC COMPANY.

The station of this company is located near the town of Newcastle, Cal., and consists of two pairs of 48-inch PELTON WHEELS, aggregating 1,200 horse-power, which run at 400 revolutions under 420 feet head.

Each pair of wheels is mounted on I beams, and direct connected to Westinghouse generators of corresponding capacity. Separate wheels run the exciters, which are also direct connected.

The water supply is furnished by the South Yuba Canal Company, an organization having over 160 miles of canals and flumes running from the summit of the Sierras down to the plains.

The power developed from this station is transmitted to Sacramento, 30 miles distant, at 15,000 volts, where it is used for lighting and general power purposes.

In the same station, a PELTON WHEEL direct connected to generator supplies Penryn, Rocklin, and other towns in the vicinity with light. The operation of this plant is in the highest degree satisfactory to all concerned.

## ONTARIO MINING COMPANY ELECTRIC STATION.

A most interesting illustration of the utilization of waste water is afforded at the Ontario Mine, Park City, Utah.

After the completion of their great drain tunnel, more than three miles in length, which taps their lode at a depth of 1,500 feet, it was found to have a continuous flow of some 1,000 cubic feet per minute. This water was piped down the cañon 1,600 feet, where a fall of 125 feet was obtained

At this point the power station is located, embracing two PELTON WHEELS connected to belt-driven electric generators. The power thus produced is transmitted six and one-half miles to the works of the company, and there used for lighting the mine, mill and hoisting works, besides furnishing power for various purposes.

The pole-line runs over a very rough country at a high elevation, subject to heavy snowfalls and electric storms of great severity, still no interruption has occurred to the service, nor have any repairs been required during the four years that it has been in operation.

## JALAPA LIGHT AND POWER COMPANY.

This station is located some twelve miles from the city of Jalapa, Mexico, and consists of two pair of 36-inch PELTON WHEELS, running under 250-foot head at 400 revolutions, direct connected to 300 K. W. Stanley three-phase generators, developing a current of 6,000 voltage, with 8,000 alternations.

The water is first taken from the river by a canal 3,500 feet long, including two tunnels aggregating 700 feet. The water supply is then brought to the station through 600 feet of 40-inch steel pipe, riveted up to form a continuous line. The pipe at one point is carried over a cañon 250 feet in depth on a steel truss bridge 10 feet wide by 100 feet in length. The receiver with which the wheels are connected is 44 inches in diameter by 60 feet long, made of $\frac{5}{16}$-inch steel plates.

The power thus generated is transmitted to Jalapa and Coatapec and is used for running a railroad system, lighting purposes, etc. Several small towns and numerous haciendas are also furnished with power and light from this station. Power is also distributed over a considerable district for driving coffee machinery.

The wheels are run in a pit outside the power-house wall, the shafts extending through same. By this means the power-house is kept thoroughly dry—a necessary condition with the high voltage of the generators. The transmission line to Jalapa is about fourteen miles in length, but the numerous branches in other directions considerably increase this distance.

## REDLANDS ELECTRIC LIGHT AND POWER COMPANY.

The station of this company is located on Mill Creek, some seven miles from Redlands, Cal., on a stream flowing from the Sierra Madre Range.

The plant, as originally installed in June, 1893, operated under a head of 340 feet, water being conveyed through a line of 30-inch pipe 7,250 feet long. The wheel plant consisted of two pair of 30-inch double-nozzle PELTON WHEELS, with capacity of 400 horse-power each, direct connected to two 250 K. W. General Electric generators, running at 600 revolutions. Also one 50-light arc generator driven by 21-inch Pelton wheel and two exciters driven by 11-inch wheels, all direct connected.

The power thus generated was carried in two circuits—one to Redlands, seven miles distant, and the other to the Union Ice Company's plant, 4½ miles distant.

After four years of successful operation, the business of the company demanded more power than was available from their water supply under the head they were then operating. To meet this demand the company extended their pipe line 3,000 feet, obtaining an effective head of 500 feet; this necessitated increasing the diameter of the wheels to conform to the same speed of the generators. New single-nozzle wheels, 34-inch diameter, were, therefore, substituted for the old ones; also a third set of wheels and generator were added; the wheels driving the exciters and arc machines were also changed to conform to the increased head.

All the wheels in this plant are regulated by relay governors, which give entire satisfaction. With this additional power, the company extended their line to Riverside, some 8 miles distant, which is now lighted from this station.

## THE OLD AND THE NEW IN HYDRAULIC PRACTICE.

From the *Scientific American*, New York.

"The Wheel illustrated on the opposite page is of the type commonly known as the Over-shot or Gravity Wheel, and is unquestionably the largest and most expensive water wheel ever constructed. It is located at Laxey, on the Isle of Man, a small island in the Irish Sea, off the west coast of England.

"This wheel is 72 ft. 6 inches in diameter, and is supposed to develop about 150 h. p., which is transmitted several hundred feet by means of wooden trussed rods having supports at regular intervals, to the bottom of which are attached small wheels running on iron ways, for purpose of lessening friction. The power thus transmitted operates a system of pumps in a lead mine, the duty of which is raising 250 gallons of water per min. an elevation of 1,200 ft. The water is brought some distance to the wheel in an underground conduit, and is carried up the masonry tower by pressure, flowing over the top into the buckets.

"This great wheel was constructed some 40 years ago, and is said to have been running continuously during all this time. It is the great attraction of the place, hundreds of visitors making the trip to the island every year to see it.

"The illustration referred to affords a very good idea of the progress made since that time in Hydraulic Engineering, and is reproduced for the purpose of showing, by way of comparison, the advantages of the modern and now generally accepted method known as the PELTON SYSTEM OF POWER.

"The little cut in the upper corner represents a PELTON WHEEL of corresponding capacity under similar conditions of head and water supply, being drawn to the same scale.

"The extraordinary results obtained from this well-known wheel are due to the peculiar shape of the buckets into which the water is directed from one or more nozzles, so that the full energy due to its head or fall is transferred into the inertia of the wheel. The power represented by the force of the water is thus converted into mechanical movement almost entirely without friction, the buckets simply taking the energy out of the stream and leaving the water inert under the wheel.

"The efficiency of the Laxey Wheel—taking resistance into account—it is estimated can not be more than 65 per cent of the theoretical power, while the PELTON will develop fully 20 per cent more, and in size and appearance is a mere toy as compared to the ponderous piece of machinery shown, with its massive column, arches, and stone foundations.

"The most striking contrast, however, will be seen in the matter of cost, which is so much less as to make a comparison almost absurd. While no data is at hand in regard to this it is apparent that it would be at least as one to fifty in favor of the PELTON.

"Such an object lesson is of value in showing the wonderful progress made in engineering practice during the last half century, in bringing the forces of nature into subjection, making them subservient to commercial and industrial purposes."

The PELTON SYSTEM OF POWER has now come into use in every civilized country on the globe, and is conceded to be one of the most useful, as well as most illustrious, inventions this country has ever produced, making possible the utilization of this great natural motive force under all conditions and for every variety of service, with a useful effect never before obtained, and in so simple a way that machinery may be said to be almost dispensed with.

THE GREAT OVER-SHOT WHEEL AT LAXEY,

ON THE ISLE OF MAN, ENGLAND.

## PELTON SYSTEM APPLIED TO PAPER MILLS.

The following is of interest, as illustrating the simple and effective method of connecting PELTON WHEELS to this class of machinery.

The Columbia River Paper Co., located at La Camas, in the state of Washington, was running five Turbines, the power from which was transmitted to the various machinery by a complicated system of counter-shafting pulleys and belting, involving a loss of power, cost of maintenance and unreliability so serious as to more than offset the advantages of such a source of power. All the Turbines were replaced in the following manner:—

To the six pulp grinders with which the mill is equipped, six double-nozzle PELTON WHEELS, 48 inches diameter, were attached by connection to driving shafts direct, without intermediate gearing. The wheels run under 120-feet head, giving the stones a speed of 250 revolutions per minute. Gates are attached to each of the wheel nozzles, operated by levers, by which means a part of the power is thrown off while the pockets are being filled, thus maintaining a uniform speed on the stones. Eight stock engines, as also two wet machines, one barker, one cut-off saw, one splitter, and a rotary pump for elevating the pulp, are all operated by a 67-inch, three-nozzle wheel, speeded at 160 revolutions.

The Fourdrinier machine is run by a PELTON WHEEL 11 feet in diameter, directly connected to the shaft of the machine, being made of this large diameter to give proper speed without gearing. This wheel runs at a regular uniform speed, meeting the exacting requirements in this way by a simple adjustment of the nozzle without governor.

An 18-inch PELTON MOTOR runs the machine shop, and a 24-inch motor a dynamo for lighting the works. With one single exception, as will be seen, all countershafts, pulleys, and belts are done away with, saving by this means fully 30 per cent in power as well as the loss of time and cost of keeping them in repair, while the output of the mill has been increased about 20 per cent.

## A JUTE MILL RUN BY ELECTRIC POWER.

An installation involving new methods of power transmission was made three years ago at Barrio Nuevo, Orizaba, Mexico. The plant consists of two 67-inch, three-nozzle PELTON WHEELS, capacity of 700 horse-power, running under 100-foot head. The wheels are connected by rope drive to four electric generators—English make—the power from which is transmitted to the factory, 1¼ miles distant.

Electric motors, varying from 1 to 20 horse-power, are attached to the various machines operated, dispensing entirely with countershafting, pulleys, and belting. This was the first equipment on so large a scale, run exclusively by electricity, with an entire absence of shafts and belt connections, and its operation has justified the most sanguine expectations of its projectors.

When it is considered that some 30 per cent of the power in any plant of this character is absorbed by shafting and belts, and that constant expense is incurred in maintaining them, the obvious advantage of such connections where electricity is the prime mover, will bring them into general use where conditions favor.

## DRAWBRIDGE OPERATED BY WATER-POWER.

The draw of the Southern Pacific Railroad Co.'s bridge over the Nechez River, at Beaumont, Texas, is operated by water-power in a somewhat novel manner. A PELTON MOTOR is located at the base of the draw and run under a 40-pound water pressure. The valve controlling the wheel is operated from the floor of the bridge by means of levers, giving the attendant absolute control of its movement. Motion is transmitted from the wheel shaft by a somewhat elaborate system of gearing, so arranged as to make the necessary reduction in speed without undue loss by friction.

Provision is made for slightly elevating the ends of the draw from their supports before turning, to facilitate the movement of so heavy a weight. The bridge has a span of 247 feet, and the weight of the draw is 41,000 pounds, the opening or closing of which requires only two minutes. The engineer of this company reports the operation of the motor as highly satisfactory, the draw being operated with the wheel running at less than half speed and without full pressure. The reliability of this means of power, as well as facility of control and small space occupied, makes it especially adapted to such service wherever it can be made available.

## DATA FOR FLUMES AND DITCHES.

To give a general idea as to the capacity of flumes and ditches for carrying water, the following data is submitted:—

The greatest safe velocity for a wooden flume is about 7 or 8 feet per second. For an earth ditch this should not exceed about 2 feet per second. In California it is the general practice to lay a flume on a grade of about ¼ inch to the rod, or often 2 inches to the 100 feet, depending on the existing conditions.

Assuming a rectangular flume 3 feet wide, running 18 inches deep, its velocity and capacity would be as shown below:—

| GRADE. | VEL. IN FT. PER SEC. | QUANTITY CU. FT. MIN. |
|---|---|---|
| ⅛ inch to rod | 2.6 | 702 |
| ¼ " " " | 3.7 | 999 |
| ½ " " " | 5.3 | 1,431 |

As the velocity of a flume or ditch is dependent largely on its size and character of formation, no more specific data than the above can be given.

It is not safe to run either ditch or flume more than about three-fourths or seven-eighths full.

## CALIFORNIA POWDER WORKS PLANT.

The plant of the above company, situated near Santa Cruz, California, illustrates the range of application of the Pelton Wheel to such an extent as to be worthy of mention.

The works are situated in a cañon, flanked by high hills, where several water sources afford moderate heads of from 30 to 50 feet. Four wheels, varying from 4 to 5 feet in diameter, are operated by this means, and drive the Chile mills, pulpers, grinders, and centrifugals used in the manufacture of powder. One of these wheels is mounted on a vertical shaft, having been substituted in the place of a turbine, which was discarded as inefficient.

The unique feature of this plant consists in the general power distribution. A 42-inch Quintex Nozzle PELTON WHEEL, operating under a head of 50 feet, drives a Dow triplex pump, forcing the water up to a reservoir situated on a hill directly above the cañon, and at an elevation of 400 feet. The water is then led back in a pipe system extending to the various departments of the works, where it is applied to PELTON MOTORS. Every machine in the entire plant is operated by a separate water motor, there being upwards of 40 used in this way. Under the head available, it requires but a small amount of water to develop a very considerable amount of power, and an endless variety of applications is afforded.

It will be readily seen that this arrangement is very economical and convenient, and eliminates all danger from fire—an all-important factor in works of this character.

## POWER FROM ARTESIAN WELLS.

Many artesian wells in various localities have a reliable flow of water under sufficient pressure to make them available for power purposes. This is especially the case in South Dakota, Nebraska, Texas, and Australia, where many wells have a pressure varying from 20 to 200 pounds per square inch. These wells have been largely utilized for power purposes by means of PELTON WHEELS, developing all the way from 10 to 100 horse-power, operating flour and feed mills, electric light plants, and various other machinery, affording a continuous and reliable power.

Where any considerable amount of power is wanted, several wells are sometimes sunk in close proximity, the united flow of which is utilized in a central station. Aside from the general adaptation of the PELTON to such conditions of service, the fact that wells under such pressure throw much sand and grit at times, makes it evident that no other wheel can be used for this purpose with any degree of reliability and satisfaction. It would be hard to conceive of a cheaper and more convenient power where obtainable by such means, and wells are now being bored extensively in localities that give a flow under sufficient pressure to make them available for power purposes.

PELTON WHEELS AND MOTORS afford facilities for utilizing these extraordinary sources of energy in the most simple, economical, and effective way. To make an intelligent estimate of the power that can be obtained from a well, it is necessary to know the diameter and depth, also the pressure while flowing through various size openings, say 1 inch, 1¼ inches, 2 inches, 2½ inches, etc. Full information will be given regarding any such application upon receipt of sufficient data.

### PUMPING PLANT PORTLAND WATER WORKS.

The new water works of the city of Portland, Oregon, though operated by the gravity system, have a pumping plant connected therewith which has some features of special interest.

Water is brought from Mount Tabor, a distance of thirty miles, in a continuous riveted steel pipe line, which terminates in a reservoir located in the city park. Below the main reservoir, which receives the flow from the pipe, is a second distributing reservoir, which gives a fall of sixty-four feet.

On the bank of this a power-house is located in which are two Reidler pumps, one of which elevates 1 000,000 gallons per day to a height of 300 feet, and the other 500,000 gallons per day to a height of 600 feet. Each of these pumps is driven by a six-foot double nozzle PELTON WHEEL connected direct to the piston rods of the pumps by crank discs and pin on the water wheel shafts. Attached to the main gate of each wheel is an automatic PELTON REGULATOR which maintains a uniform pressure on the pump discharge while varying quantities of water are being drawn off.

Two more PELTON WHEELS located in the same station furnish power to arc and incandescent dynamos for lighting the park and power-house. All the wheels are connected to a cast-iron main, thirty inches in diameter, and the water after passing the wheel runs into the lower reservoir, the power for the entire plant being obtained without any expense, and from what may be termed waste water.

The pumping station is for the purpose of supplying the elevated portions of the city which could not be reached by the main system. This plant was designed by Mr. James D. Schuyler, an engineer of large experience in work of this character, and its successful working has fully justified all expectation.

## POWER PLANT ALASKA-TREADWELL COMPANY.

The mill of the Alaska-Treadwell Gold Mining Company, located on Douglas Island, Alaska, is the largest in the world. It consists of 240 stamps, 96 concentrators, ore breakers, etc., requiring upwards of 500 horse-power. All of this machinery, covering several acres of ground, with its vast complication of shafting, pulleys and connections, is run by a single Pelton Wheel. This wheel is 7 feet in diameter, and runs under 490 feet head, at 235 revolu ions, requiring 630 cubic feet of water per minute, which is discharged through a nozzle 3¼ inches in diameter. With a 4-inch nozzle the capacity of the wheel is increased to 735 horse-power.

This company has also an 8-foot Pelton Wheel, driving a 15-drill compressor, requiring 175 horse-power, 2 reversible 5-foot wheels of 100 horse-power each for a hoist, a 6-foot wheel of same capacity running a Cornish pump, a 22-foot Pelton Wheel, running a 500-horse-power compressor, also nine other Pelton Wheels running crushers, concentrators, electric lights, etc., etc.

A new 300 stamp-mill is now under construction by this company, to be operated by Pelton Wheel 17 feet 9 inches diameter, attached direct to the battery jack-shaft—a new departure in mill practice—dispensing with counter shafts and all intermediate gearing. The crushers and concentrators are to be operated by separate wheels. All the wheels comprising this plant run under heads varying from 480 to 565 feet.

Comparisons are often made as to the relative advantage of steam and water power, and sometimes to the disparagement of the latter. With the Pelton system water has so much the better of steam that the latter is never to be thought of where water is available within any reasonable limit of cost. A notable instance of this advantage is in this case afforded.

The wheel first above referred to weighs but 800 pounds, and the entire equipment, embracing shafts, boxes, driving pulleys, etc., only about 4,000 pounds, while a steam power plant of the same capacity would not weigh less than 200 tons. The expenses of operating such a plant would be well into the thousands every month, while the cost of running Pelton Wheels is scarcely more than the oil needed for the journal bearings.

These mammoth mills, together with the various other operations of this company, afford a fair illustration of the modern methods of mining, and of how low-grade ores are made to pay large dividends.

## HIROSHIMA ELECTRIC POWER COMPANY.

This station is located at Hiroshima Falls, Japan, and consists of three 450 horse-power Pelton Wheels which run under a head of 240 feet at 300 revolutions. The wheels are 33 inches in diameter and direct connected to General Electric Generators by flexible couplings. The power developed is transmitted some five miles and operates an electric railroad and used for general power purposes. Safe and efficient regulation is afforded by electric relay governors.

## THE MANNESMANN TUBE WORKS.

This plant is situated near the town of North Adams, Mass., and consists of four 400 horse-power Pelton Wheels 11 feet 6 inches diameter, each direct connected to a train of rolls for turning out seamless tubing.

The wheels all operate under a pressure of 600 pounds, which at present is supplied by pumps. It is the intention, however, later on to operate them from a gravity system through two miles of pipe. The wheels are made of this large diameter for the purpose of giving proper speed to the machinery with which they are connected without intermediate gearing. The speed of the wheels is controlled by Pelton Governors connected to a cut-off device attached to the nozzles.

This plant has been in constant operation for two years, giving entire satisfaction.

## THE COMPAGNIE FRANCAISE HYDRAULIC.

The operations of this company—located in Trinity County, Cal.—are on a very ex-ensive scale and embrace features of special interest as regards the means of overcoming great natural obstacles in obtaining a water supply.

Three thousand inches of water are brought from Cañon Creek through flume, canal and pipe, a distance of nine miles. A pipe line 34 inches in diameter runs transversely through a cañon 200 feet in depth and 2,200 feet in length, forming a siphon. From the top of the bluff, the water is dropped through a 36-inch pipe 552 feet in length, laid on an angle of 42 degrees, securely anchored to the bed-rock.

It is then brought across a river through three pipe lines on a steel suspension bridge 5,700 feet in length. From the bridge, the pipe is carried a distance of 4,000 feet, connecting with a ditch one mile long, and the water then discharged into a reservoir, which supplies two hydraulic giants.

The laying of the pipe across the bridge and up the steep inclines was effected by means of a cable and trolley operated by a Pelton Wheel.

## NOTES REGARDING WATER RIGHTS.

It is a general principle that every owner of land upon a natural stream of water has a right to use the water for any reasonable purpose not inconsistent with a similar right in the owners of the land above, below, and opposite to him. He may take the water to supply his dwelling, to irrigate his land, or to quench the thirst of his cattle; to use it for manufacturing purposes, such as the supplying of steam boilers or the running of water wheels or other hydraulic works, so long as such use does not sensibly and injuriously affect its volume. But this is a mere privilege running with the land, not a property in the water itself.

Where the stream is small, and does not supply water more than sufficient to answer the wants of the different proprietors living on the stream, none of them can use the water for either irrigation or manufacturing, but for domestic purposes and watering stock one proprietor will be justified in consuming all the water.

Twenty years' use adverse to the right of another will give the person so using the stream the right to continue the use, regardless of the other's rights. And as to the division of water, every one who owns land situated upon a stream has the following rights: To the natural flow of the stream. That it shall continue to run in its accustomed channels. That it shall flow upon his land in its usual quantity, natural place, and usual height. That it shall flow off his land upon the land of his neighbor below, in accustomed place and at the usual level. These rights he has as an incident to the property in his land, and he can not be deprived of it by grant or location.

If any one shall make any change in the natural flow of a stream, to the material injury of any owner situated upon it, or by any interference shall prevent the stream from flowing as it was wont to flow, he is responsible for the damage he may occasion. These rights are subject to the privilege of each one to make a reasonable use of the water upon his own land while it is passing along the same. It matters not what the source of the water may be, whether it be backwater or the flowage of the same, or the water of another stream. Still the division of a stream may be made by any one, if it be returned to its natural channel before it leaves the premises.

### APPROPRIATION AND PRESCRIPTION.

No prescriptive right to the use of water of a stream can be acquired by one riparian owner as against another, by a use of the water at times when such use does not interfere with the latter's use of same, and when, as often as there is interference, the latter has protested, and sought to prevent the use. Nor is there any superiority in rights acquired in the water of a stream for the purpose of irrigating arable land over rights acquired in same for mining and milling purposes. When water of a stream has been appropriated for the purpose of running a mill, the mill owner is entitled to increase the running capacity of the mill, provided the amount of water used does not exceed the amount first appropriated. He is required to make an economic use of the water appropriated, for the purpose for which it is appropriated; and if the capacity of his ditches is greater than is necessary to provide for such use, he should be confined to the amount necessary for such economic use.—*Union Mill and Mining Company vs. Dangberg* (*81 Federal Reporter*, *73*); *United States Circuit Court.*

### NOTES REGARDING WATER RIGHTS.

It is frequently claimed that those situated at the head of a fall have certain rights and privileges over those below them. Except in peculiar cases such is not the case. For instance, a party owning all the lands on both sides of a stream, both above and below the fall, may construct a dam, and form a pond, and dispose of a certain mill site, and guarantee them certain rights in the use of all the water in the stream, should their necessities require it. He may also sell other sites with the privilege of drawing from the same pond, subject to the rights previously granted, and the party purchasing and accepting those conditions, which must be clearly specified in the deed, is bound to submit to those conditions; but other sites located upon lands below them and owned by other parties are in no way bound by such conditions as to the control of the water, but may demand the free and unrestricted use of the natural flow of the stream at all times, while those above them will be held to only a reasonable control of the water at any time.

The courts, in nearly every case where it is shown that water is used in an unreasonable manner, or diverted from its natural source, to the damage of mill owners, have promptly awarded damages for the same, and even the State has no legal right to grant the privilege of taking water from such lakes as are under State control, without the consent of the riparian owners of the lands situated upon the outlets thereof —*C. R. Tompkins, in the Modern Miller.*

The U. S. Circuit Court of Idaho, in a recent case, held that in the appropriation of water to be used at a specified place for the purpose of operating machinery and other work, and, after so using, returning to its original channel, the person so appropriating can not change the place of use to the damage of a subsequent appropriation lower down on the stream.

A late decision by the Supreme Court of Idaho confirms the doctrine, "First in time, first in right." The ground was also taken that *bona-fide* purchasers of rights of prior settlers made before the lands were surveyed, were entitled to all the rights acquired by their predecessors, even if the deeds of conveyance were defective. It was also held that an appropriator of water could convey his rights to the use thereof, and that the purchasers could use the water so acquired upon other lands than those upon which the said water was formerly used.

## NOTES REGARDING WATER SYSTEMS.

Reference has already been made to the extensive systems of ditches, flumes, and pipe lines in California, constructed for the purpose of obtaining a supply from the head waters of the various streams having their source in the very heart of the Sierras. The following data is submitted for the purpose of showing the various enterprises in this way in a single county—that of Nevada—which were constructed primarily for hydraulic mining and for power, but devoted principally now to the latter purpose These, as will be observed, aggregate a length of 489 miles, involving a cost of upwards of $3,000,000.

This estimate does not include the cost of the various pipe lines used for distribution, which would doubtless aggregate as much more. North Bloomfield ditch, 157 miles in length, cost $708,000; Milton Water Co., 80 miles, $391,000; Little York, 35 miles, $150,000; Eureka and Yuba, 63 miles, $732,000; South Yuba, 123 miles, $1,100,000, Blue Tent, 31 miles, $200,000.

Expenditures quite equal to those noted have been made in many other counties bordering on the foot-hills of the Sierra Nevada Mountains, while many similar works of even greater magnitude are projected and in process of construction in various parts of the State.

Enterprises of this character serve to indicate with what facility and profit these vast resources of power in all parts of the world can be utilized by an intelligent system of reservoirs and water-ways.

### ELECTRIC AND PNEUMATIC TRANSMISSION.

In a paper read before the Association of Engineers, of Wurtemburg, Germany, some interesting figures are given in regard to the cost of the equipment and operation of a plant of 217 horse-power, installed by the Esslinger-Cannstatt Works. The transmission in this case covered a distance of 5½ kilometers—equal to 3.4 miles—both by compressed air and by electricity. The cost of the compressed air equipment, including dam for the waterfall wheels, compressor, air motors, etc., is given as $37,500; the efficiency, 46 per cent.

The cost of the electric equipment, including dam, water-power, etc., as in the previous case, $27,500; efficiency, 69 per cent, including a loss of 15 per cent in the wiring. The cost of operation was practically the same in both cases, though only 100 horse-power was actually available with the compressed air, while 150 horse-power was delivered by the electric system, thus making the cost per horse-power delivered in the former case $37.50, and in the latter, $18.85.

The charges for depreciation were somewhat less for electricity than for compressed air, which still further favored the economy of the former.

NOTE.—The percentage of efficiency shown in either of the above examples is much below what is guaranteed by American manufacturers under similar conditions, though the relative advantages of the systems may be the same.

### MECHANICAL METHODS OF POWER TRANSMISSION.

The following table by Beringer gives the relative efficiency for different distances of the various systems indicated:—

| DISTANCE. | HYDRAULIC. | PNEUMATIC. | WIRE ROPE. |
|---|---|---|---|
| ½ mile. | 50 per cent efficiency. | 55 per cent efficiency. | 91 per cent efficiency. |
| 1 " | 49 " " | 54 " " | 85 " " |
| 3 " | 41 " " | 51 " " | 61 " " |
| 5 " | 37 " " | 50 " " | 43 " " |
| 10 " | 26 " " | 43 " " | 21 " " |
| 13 " | 18 " " | 39 " " | 11 " " |

From tne above, wire rope would appear to be much the most efficient for short distances. It may be said, with reference to the pneumatic system, that recent experiments made by William L. Saunders, C. E., of New York, in heating the compressed air at the receiving end by combustion in the mains has shown an increase of some 20 per cent in efficiency, making an excellent showing for this method.

### ELECTRICAL VERSUS MECHANICAL TRANSMISSION.

The following interesting comparison of the mechanical and electrical methods of transmitting power was given by Mr. L. B. Stillwell at a meeting of railway men: A steel cable 1½ inches in diameter, traveling at the rate of twelve miles per hour, can transmit nearly 2,000 horse-power. But by taking a copper wire 1 square inch in section and applying it to a potential equal to that which is in use to-day in at least one place in this country, viz., 10,000 volts, at 1,000 amperes per square inch, we find we are transmitting in an invisible form over that wire more than 13,000 horse-power, which is enough to rupture instantly six such cables as are ordinarily used in operating a cable railway. As much power can be transmitted through such a copper wire, under the conditions named, as through six such belts as were seen at the World's Fair, six feet wide, and running at the rate of a mile per minute.

## PELTON WATER WHEELS IN THE NAVAL SERVICE.

The following is an abstract from a very elaborate and exhaustive report to the Secretary of the Navy, by Lieut. F. J. Haeseler, of the Naval Academy at Annapolis, on the adoption of PELTON WHEELS—operated by steam pumps—for auxiliary power purposes on board vessels of war.

"The water motor used in the before described comparisons is manufactured by the PELTON WATER WHEEL COMPANY, San Francisco and New York. I have tested other makes of water wheels, and read of still other tests, and it has been my experience, as well as that of others making comparative tests of these wheels, that they are at least 15 per cent, and in some cases 35 per cent, more efficient. The firm claims an efficiency of 85 per cent when the wheels are set in accordance with their instructions, and I have found their claims not only true, but below what is really obtainable.

"During the past two months, Ensign W. H. G. Bullard, U. S. Navy, and the writer, tested one of the PELTON WHEELS bought out of stock two years ago, and without any expectations at that time of its being tested for efficiency. Tabulated below are the results of our tests, showing an efficiency at the higher pressures in excess of that claimed by the firm.

TABULATED TESTS OF A PELTON WATER WHEEL MADE AT THE U. S. NAVAL ACADEMY.

| Test No. | Size jet. | Running pressure. | Revolutions. | Actual water used. | Actual h. p. developed. | Theoretical h. p. possible. | Efficiency percentage. |
|---|---|---|---|---|---|---|---|
| 1 | [illegible] in. | 90 | 775 | 14.05 | 4.101 | 5.496 | 74.60 x |
| 2 | [illegible] in. | 105 | 910 | 15.14 | 5.025 | 6.910 | 72.70 x |
| 3 | [illegible] in. | 100 | 850 | 14.78 | 4.694 | 6.422 | 73.10 x |
| 4 | [illegible] in. | 100 | 775 | 11.78 | 5.349 | 6.422 | 83.30 |
| 5 | [illegible] in. | 103 | 780 | 15.00 | 5.563 | 6.715 | 82.90 |
| 6 | [illegible] in. | 125 | 880 | 23.05 | 10.730 | 12.520 | 85.69 |
| 7 | [illegible] in. | 102 | 775 | 20.82 | 7.845 | 9.226 | 85.02 |
| 8 | [illegible] in. | 100 | 775 | 20.61 | 7.756 | 8.957 | 86.59 |
| 9 | [illegible] in. | 125 | 900 | 23.05 | 10.670 | 12.520 | 85.16 |
| 10 | [illegible] in. | 73 | 775 | 17.61 | 4.457 | 5.587 | 79.79 x |
| 11 | [illegible] in. | 86 | 880 | 19.11 | 5.365 | 7.143 | 75.12 x |
| 12 | [illegible] in. | 100 | 780 | 20.61 | 7.717 | 8.957 | 86.15 |

NOTE.—Tests marked thus x were made to find the efficiency of the governor. The wheel was running at incorrect number of revolutions at the time for the pressure at the jet.

"Referring to the foregoing tests, the amount of water used was absolutely measured by running it into the Natatorium of the Naval Academy. The scale, break, and other appliances used in the tests were most carefully adjusted, and every care used to insure perfect accuracy. The efficiencies thus determined, 86 per cent at full load, and 82⅓ per cent at half load, can therefore be considered reliable, and agree with those in testimonials published by the PELTON firm. In this connection it may be stated that the PELTON COMPANY claim in many cases efficiencies running considerably higher than those here obtained, and refer to various working tests made in California, running from 87 to 92 per cent.

"The experience of the users of PELTON WHEELS shows that absolutely no repairs are required for an indefinite time, and no attendance other than filling the oil cups once a day. A water motor has a wider range of power than any steam engine. With the same PELTON WHEEL I have got 16 horse power at an efficiency of 86 per cent, and ¾ horse power at 82 per cent efficiency. Where is there a steam engine that can equal that performance? This same motor is capable of developing at least 60 horse power at the same high efficiency, it being simply a question of water supply and pressure in any given case to meet all requirements.

[Signed] LIEUT. F. J. HAESELER, U. S. N."

## SOME FEW REFERENCES ON PELTON WHEELS,

### OF CAPACITIES VARYING FROM 20 TO 1,500 H P. EACH.

SEVERAL THOUSAND ADDITIONAL COULD BE GIVEN, BUT THESE ARE SUFFICIENT TO INDICATE THE EXTENT AND VARIETY OF SERVICE AND THE REPUTABLE CHARACTER OF PATRONAGE.

Idaho Mine, Grass Valley, Cal....................18
Empire Mine, Grass Valley, Cal.................. 6
Crown Point Mine, Grass Valley, Cal........... 2
Maryland Mine, Grass Valley, Cal.............. 1
North Star Mine, Grass Valley, Cal.............10
Badger Mine, Grass Valley, Cal................... 1
Andrews Bros.' Mill, Grass Valley, Cal......... 1
New York Hill Mine, Grass Valley, Cal....... 1
Binkleman's Brewery, Grass Valley, Cal. . ...... 1
S. D. Avery, Grass Valley, Cal..................... 1
Providence Mine, Nevada City, Cal.............. 3
Merrifield M., Nev. City, (displacing Knight's) 1
Wyoming Mine, Nevada Co., Cal................. 3
Nevada City Mine, Nevada Co., Cal ............ 5
Charonnat Mine, Nevada Co., Cal................. 4
Mayflower Mine, Nevada Co., Cal................. 2
Ella Mine, Nevada Co., Cal........................ 1
Little Dublin Mine, Nevada Co., Cal ........... 1
Greenman Mine, Nevada Co., Cal................ 1
Nevada County Mine, Nevada, Cal............... 1
Daily Herald Pub. Co., Nevada Co., Cal....... 1
James J. Ott, Nevada Co., Cal.................... 1
Pittsburgh Mine, Nevada Co., Cal.............. 4
North Banner Mine, Nevada Co., Cal........... 5
Maltman's Works, Nevada City, Cal............ 1
Spargo Mine, Nevada Co., Cal...................... 1
Boss Mine, North San Juan, Cal.................. 1
Eagle Bird Mine, Washington, Cal.............. 6
Yuba Mine, Washington, Cal....................... 2
Spanish Mine, Washington, Cal................... 1
Grafton Mine, Washington, Cal................... 1
Cornucopia Mine, Washington, Cal.............. 1
San Jose Mine, Washington, Cal.................. 1
Rocky Glen Mine, Graniteville, Cal......... ..... 3
Eureka Lake Saw Mill, Columbia Hill, Cal..... 1
Delhi Mine, Columbia Hill, Cal.................... 3
Derbec Mine, North Bloomfield, Cal............. 1
Helwig & Co., North Bloomfield, Cal........... 1
Sierra Buttes M. Sierra City (displac'g Knight's) 4
Young America Mine, Sierra City, Cal.......... 4
Cleveland Mine, Sierra City, Cal.................. 1
Goff Mine, Sierra City, Cal.......................... 1
Mooney Mine, Sierra City, Cal ..................... 1
Wixon Mine, Sierra City, Cal....................... 1
Sietz Mine, Sierra City, Cal......................... 1
Sierra Phœnix, Poker Flat, Cal.................... 1
Belmont Consolidated Mine, Poker Flat, Cal.. 1
Rainbow Mine, Alleghany, Cal..................... 3
Oceola Mine, Alleghany, Cal...................... 1
Home Stake Mine, Forest City, Cal.............. 1
Bald Mount'n Extens'n Mine, Forest City, Cal 1
H. K. Turner's Saw Mill, Sierra Valley, Cal... 2
J. and J. Blair, Placerville, Cal................... 1
Cook a Sibeck's Flour Mill, Placerville, Cal.... 1
Rulison & Likens' Mine, Placerville, Cal....... 1
Mentone Stone Works, Redlands, Cal........... 1
Oakland Gold Mining Co., El Dorado Co., Cal. 3
Horman Mine, Placer Co., Cal...................... 1
Michigan Mining Co., Valley Springs, Cal...... 1
Western Beet Sugar Co., Watsonville, Cal...... 2
Crystal Mine, Shingle Springs, Cal.............. 2
Calaveras Con. Gold Mining Co., Angels..... 2
Gold Hill Mining Co., Angels, Cal............ 1

Treadwell Mill, Douglas Island, Alaska, (displacing Knight's)........................ 8
Alaska Mill & Mining Co., Douglas Is., Alaska 3
J. P. Gross, Shingle Springs, Cal................ 1
Roberts & Wickham Mine, Garden Valley, Cal 3
Gold Blossom Mine, Ophir, Cal..................... 2
Belmont Mine, Ophir, Cal........................... 1
Boulder Mine, Newcastle, Cal...................... 2
Dam Mine, Michigan Bluffs, Cal.................. 1
Mayflower Mine, Forest Hill, Cal..... ........... 3
Golden Eagle Mine, Butte Co., Cal............... 1
Zeile Mine, Jackson, Cal. (displacing Knight's) 3
Mammoth Mine, Jackson, Cal...................... 1
Kennedy Mine, Jackson, Cal., (displacing Knight's) 7
Plymouth Mine, Plymouth, Cal..................... 7
Bandereta Mine, Coulterville, Cal................ 1
Oro Plata Mine, Murphy's Cal...................... 1
Morgan Mine, Carson Hill, Cal..................... 1
Comet Mine, San Andreas, Cal..................... 1
Union Mine, San Andreas, Cal..................... 2
Francis Mine, Hornitos, Cal........................ 1
Casy Mine, Bishop Creek, Cal..................... 1
Los Gatos Flour Mill, Los Gatos, Cal............. 2
Quartz Hill Mine, Scott's Bar, Cal............... 1
Fine Gold Mine, Valley Springs, Cal............ 1
Cal. Sugar Refinery, San Francisco, Cal........ 2
Barnhardt & Bros., Downieville, Cal.............. 1
Carney, Mine, Downieville, Cal.................... 1
York, or Slug Gulch Mine, Downieville, Cal.. 1
Butcher Ranch Mine, Downieville, Cal......... 1
Milton Mine, Grizzly Flat, Cal..................... 1
Gopher Boulder Mine, Kelsey, Cal............... 1
Burnham Mine, Georgetown, Cal.................. 1
Buck Mine, Moore's Flat, Cal....................... 1
Brand City Mine, Brandy City, Cal.............. 4
Green Mountain Mine, Crescent Mills, Cal..... 2
Crescent Mine, Crescent Mills, Cal.............. 3
Taylor Plumas Mine, Crescent Mills, Cal....... 1
Premium Mine, Crescent Mills, Cal ............. 1
Indian Valley Mine, Greenville, Cal........... 2
Arcadia Mine, Greenville, Cal........... ....... 1
Flour Mill, Greenville, Cal......................... 1
Knickrum & Sutton Sawmill, Mohawk Val. Cal. 1
Black Bear Mine, Siskiyou Co., Cal............. 1
Klamath Mills Mine, Siskiyou, Co., Cal......... 1
Quartz Mountain Mine, Fresno, Cal............. 1
Mountain Tunnel Gravel Mine, Todd's Val. Cal. 1
J. G. Tolten & Co. Sawmill, Beckwith, Val. Cal. 1
Pioneer Pulp Mill, Alta, Cal........................ 5
J. S. Landis, Smartsville, Cal..... ................. 1
P. A. Chalfant, Independence, Cal....... ......... 1
R. B. Lane, Stockton, Cal...... ..................... 1
Gold Canyon Mine, Moore's Flat, Cal........... 5
Empire Mine, Downieville, Cal.................... 1
Phenix Mine, Sierra City, Cal..................... [illegible]
Esmeralda Mine, Murphy's, Cal.................... 1
Quicksilver Mining Co., New Almaden, Cal... 1
Kanaka Mine, Groveland P. O., Cal.............. 1
M. B. Kelley, Pulp Mill, Lower Cascades, Or. 1
Foster Mine, Ashcroft, British Columbia........ 2
The Quail Gold Mine, Mariposa Co., Cal....... 1
Binet Bros., Clipper Mills, Butte Co., Cal... 1
William Metcalf, Nelson Point, Cal............ 1

## SOME UNITED STATES REFERENCES.

Amador Gold Mining Co., Jackson, Cal. ......10
Garfield & Hayes Sawmill, Sierra City, Cal.... 1
L. L. Blatchley, Sawmill, Etta, Cal.............. 1
Hartman-Taberin Mine, Shingle Springs, Cal.. 1
O. W. McKenzie, Ione, Cal......................... 1
J. C. Kerickrem, Mohawk, Cal......... .......... 1
Live Oak Mine, Iowa Hill, Cal...................... 1
Penn Chemical Works, Valley Springs, Cal... 2
Mountaineer Mine, Nevada City, Cal........... 1
St. John's Mine, Nevada City, Cal................ 1
M. C. Taylor, Grass Valley, Cal................. 1
Gen. Grant Mine, Pike City, Cal.................. 1
Uncle Sam Mine, Squaw Creek, Shasta Co., Cal. 1
Wm. Hopkins, Redwood City, Cal............... 1
Reiley & Bliss Mine, Copley, Cal................ 1
A. Leary, Sierra City, Cal............................ 1
Coe Mine, Grass Valley, Cal........................ 3
American Eagle Mining Co., Butte Co., Cal.... 1
Heath Mining Co., Baker City, Oregon.......... 1
A. J. Atwell, Visalia, Cal ........................ 1
Taylor & Forbes, Downieville, Sierra, Co., Cal. 4
Last Chance Mining Co., North Bloomfield, Cal. 2
Champion Mining Co., Nevada City, Cal........ 3
Watrous Mine, Baker City, Oregon............... 1
John Scott, Sierra City, Cal........................ 1
Brunswick Mining Co., Grass Valley, Cal...... 1
Mercer & Salinas Mining Co., Sierra City, Cal. 1
Alpha Mining Co., Grass Valley, Cal............ 3
Daisy Cement Mine, Washington, Cal........... 1
Nevada Electric Light Co., Grass Valley, Cal.. 1
Nevada City Electric Light Co., Nevada, Cal. 1
Keystone Mine, Sierra City, Cal.................. 1
Equator Mining Co., Dimond Springs, Cal...... 1
Omaha & Lone Jack Mining Co., Grass Val. 3
Wheeler Mill, French Gulch, Shasta Co., Cal. 1
Pet Hill Mining Co., Grass Valley Cal.......... 1
Wilson Mining Co., Baker City, Oregon......... 1
Dardanelles Mining Co., Forest Hill, Placer Co. 1
Big Bend Tunnel & Mining Co., Butte Co. Cal. 1
Arrow Head Springs Hotel, San Bernardino... 1
Quaker Mine, Calaveras Co., Cal.................. 1
Church Mining Co., El Dorado, Cal............... 3
Alpine Mining Co., Georgetown, Cal............ 2
Golden Chest Mining Co., Trinity Co., Cal...... 1
Vallecito Gold Mining Co., Calaveras Co., Cal. 1
Kentuck Mine, Shoup, Idaho......................... 1
Parker Mining Co., Ketchum, Idaho............. 1
A. A. Pollard Mill, Silver City, Nevada......... 2
C. C. Stevenson, Mill & Hoist, Gold Hill, Nev. 2
Beaver Head Hydraulic M. Co., Salmon C'y, Id. 2
Fulton Foundry Mill, Gold Hill, Nevada...... 1
Ramshorn Mine, Bay Horse, Idaho ............. 2
Virginia City Electric Light Co., Vir. City, Nev. 4
G. Riebold, Warrens, Idaho......................... 1
Monte Cristo Mine, Mono Co., Cal................ 1
Excelsior & Eureka Mines, Baker City, Or... 1
Superior Mining Co., Placerville, Cal........... 2
Nevada M. & M. Co., Virg. City, Nevada........ 8
California & Con. Virg. Mill, Virg. City, Nev. 6
Star Mining & Reduction Co., Fresno Co., Cal. 1
Eureka Con. Mining Co., Amador Co., Cal...... 2
Armstrong Mine, Golden, B. C..................... 1
Excelsior Water & Mining Co., Smartsville, Cl. 1
Bunker Hill Mining Co., Cœur d'Alene, Idaho. 1
Heath Mine, Ruthburg, Idaho...................... 1
Borger Mine, Calaveras Co., Cal.................. 1
Gold Ring Mine, Placer Co., Cal........... ..... 1
Champion Mining Co., Nevada City, Cal..... 1
Idlewild Gold Mining Co., Greenwood, Cal.. 5
Development Syndicate, Butte Co., Cal....... 2

Banner Mine, Calaveras Co., Cal .................. 1
Bunker Hill & Sullivan Mine, Idaho............ 6
Sunnyside Mine, Tuolumne Co., Cal............. 1
Homestake Mine, Custer Co., Idaho............. 1
Ilex Mining Co., Mokelumne Hill, Cal.......... 5
Arroyo Seco Mine, Ione, Cal. ...................... 1
Truckee, "Republican," Truckee, Cal........... 1
Mammoth Bar Gold Mining Co., Auburn, Cal. 2
Calaveras Blue Gravel M. Co., Mokelumne, Cal. 1
Buffalo Con. Mining Co., Sierra City, Cal...... 1
Conner Creek M. & M. Co., Conner Creek, Or. 1
Newhall Times, Burbank, Cal...................... 1
Golden Queen Mine, Forbestown, Butte Co. Cal, 2
Hathaway Mine, Newcastle, Cal.................... 2
Gold Bank Mine, Forbestown, Butte Co., Cal. 5
Platt & Gilson Mining Co., Soulsbyville, Cal. 2
Utica Mine, Angel's Camp, Cal..................... 5
Pittsburg Mining Co., Nevada Co., Cal......... 4
Placer Co. Angus, Auburn, Cal..................... 1
Rawson Sawmill, Bishop Creek, Cal... ........ 1
Moreley Mining Co., Calaveras Co., Cal........ 3
Calaveras Con. Mining Co., Calaveras Co., Cal. 4
Big Sandy Mine, El Dorado Co., Cal............ 1
Linden Gravel Mine, El Dorado, Cal............ 1
Harrison Mine, Butte Co., Cal..................... 1
North Alaska Mine, Sierra Co., Cal............... 1
Bandarita Mine, Mariposa Co., Cal........ ...... 1
Diamond Creek Con. Mining Co. Nev. Co. Cal. 6
Yuba Mine, Maybert, Nevada Co., Cal........ 1
Sisson Lumber Co., Siskiyou Co., Cal........... 1
Golden Gate Mine, Tuolumne Co., Cal......... 3
Tiger Mining Co., Burke, Idaho.................... 7
Gigmag Gravel Mining Co., Placerville, Cal.. 1
North Star Mining Co., Trinity Co., Cal........ 1
Saratoga Water Works, Saratoga, Cal............ 1
Plymouth Rock Mine, Calaveras Co., Cal...... 1
Emma & Last Chance Mine, Wardner, Idaho.. 2
Enterprise Mine, Weaverville, Cal................ 1
Giant Powder Co., Clipper Gap, Cal........... 6
Electric Light & Power W'ks, Watsonville, Cal. 1
Hog's Back Mine, Placer Co., Cal................ 1
Diadem Mine, Meadow Valley, Cal............... 1
Idle Wild Mine, El Dorado Co., Cal.............. 2
Eureka Gold Mining Co., Calaveras Co., Cal. 2
American River Syndicate, El Dorado Co. Cal. 1
Cosmopolitan Mining Co., Amador Co., Cal..... 3
H. J. Lewelling, Elec. Light W'ks, St. Helena. 1
Belshaw & Hayes Mine, El Dorado, Co. Cal. 1
Reveille Printing Works, Baker City, Or........ 1
Marysville Democrat, Marysville, Cal........... 1
Beach & Co., Planing Mills, Placerville, Cal... 1
Sierra Nev. Land & Water Impt. Co., Plac'v'le 4
W. P. Oliver Machine Shop, Grass Val., Cal... 1
Otto Norman, Howell Mountain, Napa Co., Cal. 1
F. M. Steel, Pescadero, Cal.......................... 1
Hurley G. M. Co., Valley Springs, Cal.......... 2
Smith's Sawmill, East Fork, Trinity Co., Cal.. 1
Mount Charles Gravel M'g Co., Nev. Co., Cal. 1
Blue Point Gravel M'g Co., Smartsville, Cal. 1
Ship Mine, Sierra Co., Cal............... ......... .
Gold Bluff Mine, Sierra Co., Cal................ . 1
Keystone Mine, Sierra Co., Cal..................... 1
Eagle G. M. Co., Nelson, B. C...................... 1
Spanish Mine, Nevada Co., Cal..................... 1
The R. C. Mine, Brownville, Yuba Co., Cal...... 1
Ayer Mine, Rooney Flat, Cal........................ 1
Ione Trade School, Ione, Cal........................ 4
Silver Grey Mine, Detrick, Cal ..... ......... 1
Gwinn Mine, Calaveras Co , Cal.. ... ....... 2
Sonora Electric Light Co., Sonora, Cal......... 1

## SOME UNITED STATES REFERENCES.

| Reference | No. |
|---|---|
| Roaring Fork Electric Light & Power Co., Aspen, Colo (displacing turbines) | 15 |
| Aspen Mining & Smelting Co., Aspen, Colo | 4 |
| Veteran Tunnel Co., Aspen, Colo | 1 |
| Aspen Mining Co., Aspen, Colo | 4 |
| Castle Creek Tunnel & Power Co., Aspen, Colo | 6 |
| Holden Smelting & Milling Co., Aspen, Colo | 2 |
| Ruby Trust Mining Co., Ouray, Colo | 3 |
| Revenue Tunnel Co., Ouray, Colo | 1 |
| Amador Lilla Mining Co., Ouray, Colo | 1 |
| Michael Breen Mining Co., Ouray, Colo | 3 |
| Caroline Mining Co., Ouray, Colo | 2 |
| Kansas City & Ouray Mining Co., Ouray, Colo | 2 |
| Hector Gold Mining Co., Ouray, Colo | 1 |
| Mineral Farm M. & M. Co., Ouray, Colo | 1 |
| Sheridan & Mendota Mill, Telluride, Colo | 3 |
| Gold King Mining Co., Telluride, Colo | 2 |
| Glenwood Light & Power Co., Glenwood Springs, Colo | 1 |
| Bennett Saw & Planing Mill, Glenwood Springs, Colo | 1 |
| North Star Mining Co., Silverton, Colo | 2 |
| Silver Lake Mines, Silverton, Colo | 1 |
| Prussian Mines, Boulder, Colorado | 1 |
| Argentum Juniata Mining Co., Aspen, Colo | 1 |
| Mt. Wilson G. & S. M. Co., Seymour, Colo | 1 |
| Smuggler Mining Co., Aspen, Colo | 1 |
| Hector Mine, Ouray, Colorado | 1 |
| Pueblo S. & R. Co., Pueblo, Colorado | 1 |
| Bimetallic M. & M. Co., Aspen, Colorado | 2 |
| Mollie Gibson M. & M. Co., Aspen, Colo | 3 |
| Thomas Stephens, Montezuma, Colorado | 2 |
| James H. McCoy, Ouray, Colorado | 1 |
| Suffolk M. & M. Company, Ophir, Colo | 2 |
| Magna Charta S. M. Co., Tomichi, Colo | 1 |
| French Mountain G. M. Co., Gold Park, Colo | 1 |
| Lincoln Park Irrigating Co., Canon City, Colo | 1 |
| D. C. Hartwell, Ridgeway, Colo | 1 |
| J. B. Hudson, Gardiner, Colo | 1 |
| Electric Light Co., Crested Buttes, Colo | 1 |
| St. Louis & Colorado S. & M. Co., Thomasville Colo | 2 |
| Parrott Silver & Copper Co., Butte, Montana | 1 |
| Black Oak Mining Co., Meagher Co., Montana | 1 |
| Mintzer Mine, Red Rock, Red Rock, Montana | 1 |
| Anaconda Smelting Works, Butte, Montana | 1 |
| Golden Messenger Mine, York, Montana | 1 |
| J. E. Woods, Silver Star, Montana | 1 |
| Bimetallic Mining Co., Phillipsburg, Mont | 2 |
| Tuttle Mfg. & Supply Co., Anaconda, Mont. | 3 |
| W. C. Miller, Dillon, Montana | 1 |
| Hope Mining Co., Phillipsburg, Mont | 1 |
| Boise Iron and Red. Works, Boise, Idaho | 1 |
| Elmira Silver M. Co., Banner, Idaho | 1 |
| Morning Mine, Mullan, Idaho | 2 |
| James Moore, Weiser, Idaho | 1 |
| Gem Mining Co., Gem, Idaho | 1 |
| De Lamar Mining Co., De Lamar, Idaho | 2 |
| Gibbonsville G. M. & M. Co., Idaho | 1 |
| Milwaukee Mining Co., Wallace, Idaho | 2 |
| Bunker Hill & Sul. M. Co., Wardner, Idaho | 1 |
| De Lamar Tailing Mill, De Lamar, Idaho | 2 |
| Lemhi Placer M. Co., Salmon City, Idaho | 1 |
| C. L. Coleman, Blackfoot, Idaho | 1 |
| Columbia Con. G. M. Co., Challis, Idaho | 1 |
| Yellow Jacket Mining Co., Idaho | 1 |
| Niobrara Mill Co., Niobrara, Neb | 1 |
| North Star Mine, Grass Valley, Cal | 4 |
| Grass Valley Electric Light Co., Nevada Co. | 1 |
| Brown Bear Mining Co., Trinity Co., Cal | 1 |
| McEwen & Marquis Mining Co., Telluride, Colo. | 1 |
| Beattie & McWilliams, Telluride, Colo | 2 |
| Belmont Con. Mining Co., Telluride, Colo | 1 |
| Weller, Jones & Weller, Telluride, Colo | 1 |
| Iron Silver Mining Co., Leadville, Colo | 1 |
| Silver Cord Com. Mining Co., Leadville, Colo | 1 |
| Gilman Electric Lt. & Power Plant, Gilman, Colo | 1 |
| Commonwealth Mining Co., Idaho Springs Colo. | 3 |
| John Rudberg, Idaho Springs, Colo | 1 |
| Colorado Central Mining Co., Georgetown, Colo. | 2 |
| Hall's Sampling Works, Georgetown, Colo | 1 |
| Georgetown Electric Light & Power Co., Georgetown, Colo | 1 |
| Pandora & Oriental Mining Co., Montrose, Colo. | 1 |
| S. S. Metzler, Irwin, Colo | 1 |
| Colorado Land & Improvement Co., Glenwood Springs, Colo | 4 |
| Mohawk Mining Co., Breckinridge, Colo | 1 |
| Lewis Mining Co., Silverton, Colo | 1 |
| Rico Reduction Works, Rico, Colo | 1 |
| Terry Milling Co., Eureka, Colo | 2 |
| People's Lt. & P. Co., Aspen, Colo | 2 |
| San Miguel Con. M. Co., Telluride, Colo | 1 |
| A. E. Reynolds, Ouray, Colorado | 1 |
| Golden King M. Co., Alleghany, Colo | 1 |
| Carribeau & Mon. Mines, Ophir, Colo | 1 |
| St. Vrain W. & P. Co., Boulder, Colo | 1 |
| Colorado Iron Works, Denver, Colo | 12 |
| Orangeville Roller Mill Co., Price, Utah | 1 |
| Ophir Hill M. & Con. Co., Ophir, Utah | 2 |
| Butterfield Mining Co., Bingham, Utah | 1 |
| Mayfield Milling Co, Mayfield, Utah | 1 |
| R. M. Jones, Salt Lake City, Utah | 1 |
| Mt. Pleasant R. Mills, Mt. Pleasant, Utah | 1 |
| Bozeman Electric Light Co., Bozeman, Montana (displacing turbine) | 1 |
| Ontario Silver Mining Co., Park City Utah | 1 |
| Daly Mining Co., Park City, Utah | 1 |
| Union Concetrating Co., Park City, Utah | 1 |
| Utah Gem Mine, Alpine, Utah, Displ. Turbine | 1 |
| Rodgers Milling Co., Bingham Cañon, Utah | 1 |
| George M. Scott Co., Salt Lake City, Utah | 1 |
| Utah & Montana Machinery Co., Utah | 2 |
| Manti Roller Mills, Manti, Utah (displacing turbine) | 1 |
| John Southworth, Tooele City, Utah | 1 |
| Silverton Mining Co., Silverton, Colo | 1 |
| Hutterische Society, Bon Homme, S. Dakota | 1 |
| Underwood Milling Co., Northville, S. Dak. | 1 |
| Armour Roller Mills, Armour, S. Dakota | 1 |
| Excelsior Mill Co., Yankton, S. Dakota | 1 |
| Woonsocket R. Mills, Woonsocket, S. Dak | 1 |
| Carson River Pl. M. & D. Co., Dayton, Nev. | 1 |
| E. Williams, Empire City, Nevada | 1 |
| Cook & Hamlin, Hawthorne, Nevada | 1 |
| Commercial Mining Co., Prescott, A. T. | 1 |
| Brown & Martin, Silver City, New Mexico | 1 |
| Wanoosnock El. Power Co., Fitchburg, Mass. | 6 |
| Deerfield Valley Mills, Charlemont, Mass | 1 |
| Frary Mfg. Company, Charlemont, N. H. | 1 |
| M. C. Wentworth's Hotel, Jackson, N. H. | 1 |
| Colebrook El. Light & Power Co., N. H. | 1 |
| O. W. Norcross, Marble Hill, Ga | 1 |
| Mfg. Investment Co., Appleton, Wis | 1 |
| Penn. Railroad Co., Summit Branch, Penn | |
| John H. Miller, Tyrone, Penn | |
| S. Pacific Railway Co., Beaumont, Texas | |
| Silver State Mining Co., Nevada | |
| Valley View Mine, Placer Co., Cal | |
| Yellowstone Mining Co., Trinity Co., Cal | |

## SOME UNITED STATES REFERENCES.

Columbia River Paper Co., La Camas, Wash. (displacing turbines) ... 10
Washington Stone Co., Spokane Falls, Wash... 1
Cascade Bay L. & M. Co., Newhall, Wash... 1
Fidelity Trust Co., Tacoma, Wash... 1
Holley, Mason & Marks, Spokane, Wash... 1
Holley, Mason & Marks, Coulee City, Wash... 1
Bunker Hill Mining Co., Shasta Co., Cal... 1
Eclipse Gold Mining Co., Placerville, Cal... 2
Toll Gate Mine, El Dorado Co., Cal... 1
Riverside Mine, Calaveras Co., Cal... 1
Lone Star Mine, Calaveras Co., Cal... 1
I. X. L. Mine, Nevada Co., Cal... 3
Gem Mining Co., Redding, Shasta Co., Cal... 1
McKillican M. & M. Co., North Bloomfield, Cal. 2
Auburn Ice Works, Auburn, Cal... 1
Roll Kennison, Auburn, Cal... 1
Sipp Mining Co., Blue Cañon, Placer Co., Cal... 1
San Juan Mining Co., Placer Co., Cal... 1
Payne's Saw Mill, Callahan's Ranch, Cal., (displacing turbine)... 1
Gover Mining Co., Amador City, Cal... 2
Clark Mill, Williams, Colusa Co., Cal... 1
Excelsior Mining Co., Smartsville, Cal... 1
Mina Rica Mining Co., Auburn, Cal... 1
Azusa Ice & Cold Storage Co., Azusa, Cal... 1
Oak Con. Mining Co., Grizzly Flat, Cal... 2
Bonanza Mill & Mining Co., Sonora, Cal... 2
Wurmser Merrall Placer M'ng Co., Newhall, Cal 2
Golden Era Mine, Columbia, Cal... 1
Colombo Mine, Sierra City, Cal. (disp. Knights) 1
Phœnix Mill & Mining Co., Sierra City, Cal... 1
Fegundes Mine, Gazelle, Cal... 1
Wilderness Gold Mining Co., Plumas Co., Cal... 1
Mahoney Mining Co., Sutter Creek, Cal... 1
Jameson Mining Co., Johnsville, Cal... 2
North Star Lumber Mill, Dunsmuir, Cal... 1
Cuyamaca Mill & Mining Co., San Diego Co., Cal 1
Noble Quartz Mill, San Diego, Cal... 1
Empire Mining Co., Siskiyou Co., Cal... 1
Parmelee Ore Con. Co., Copley, Cal... 1
Washington Mine, Sheep Ranch, Cal... 1
Gladstone Mining Co., French Gulch, Cal... 2
Lick Observatory, Mt. Hamilton, Cal... 1
University of California, Berkeley, Cal... 3
Union Con. Drift Mining Co., Marysville, Cal... 1
Bucks Ranch Mining Co., Weaverville, Cal... 1
Boston Mine, Nevada Co., Cal... 1
The Short Mine, Siskiyou Co., Cal... 1
Pine Valley Mine, San Diego Co., Cal... 1
Bear Valley Mine, Mariposa Co., Cal... 1
Norfolk Mine, Calaveras Co., Cal... 1
National Mine, Eureka, Nevada Co., Cal... 6
Gold Ball Mining Co., Siskiyou Co., Cal... 1
Mercer Con. G. M. Co., Sierra Co., Cal... 1
Chips Mining Co., Sierra Co., Cal... 1
Butcher's Ranch M. Co., Sierra Co., Cal... 1
Rainton Mine, Sierra Co., California ... 1
Chloride Mining Co., Trinity Co., Cal... 1
Titusville Mill, San Juan, California... 1
Mountain Ledge Mine, Trinity Co., Cal... 1
Van Mining Co., Georgetown, Cal ... 1
Independence Mine, Placer Co., Cal... 1
Harrison Mine, Plumas Co., Cal... 1
Turner Sawmill, Sierra Valley, Cal ... 1
Almaden Mines, New Almaden, Cal... 2
San Antonio L. & P. Co., Pomona, Cal... 4
Leland Stanford Vina, Tehama Co., Cal... 1
Gold Band Mine, Forbestown, Cal... 4
Standard Con. Mining Co., Bodie, Cal... 4
Hector Gold Mining Co., Sutter Creek, Cal. 1
Mountain Ledge Co., Sierra City, Cal... 2
Diamond Creek Con. Min. Co., Nevada Co., Cal 1
Lake Hemet Water Co., Los Angeles Co., Cal... 2
Bender Mining Co., Siskiyou Co., Cal... 1
Clinton Con. Mining Co., Calaveras Co., Cal... 1
Ontario Land & Improvement Co., Ontario, Cal. 1
Gold Bar Mining Co., Etna, Cal... 1
Valley View Mining Co., Lincoln, Cal... 1
Bell Electric Co., Auburn, Cal. (displacing turbine) 1
Union Shaft Placer M. Co., Valley Springs, Cal 1
Casco Mill & Mining Co., Amador Co., Cal... 1
The Geysers Electric Lt. Station, Geysers, Cal. 1
Eclipse Mining Co., Butte Co., Cal... 1
Shasta Mineral Water Co., Shasta Springs, Cal. 1
San Diego Land & Town Co., National City, Cal 1
Los Gatos Ice Co., Los Gatos, Cal... 1
Odin Gold & Silver M. Co., Calaveras Co., Cal.. 1
Bear Valley Water Works, Redlands, Cal... 1
New York Mine, Amador Co., Cal... 1
Jackson & Lakeview Mining Co., Lundy, Cal... 1
Gold Flat Mining Co., Nevada Co., Cal... 1
Harper & Reynolds, Los Angeles, Cal... 1
Occidental Con. Mining Co., Virginia, Nev... 1
Pioche Con. Mining Co., Pioche, Nev... 1
Black Bear Mining Co., Wallace, Idaho... 2
Custer Mining Co., Wallace, Idaho... 1
Granite Mining Co., Wardner, Idaho... 1
Lemhi Mining Co., Salmon City, Idaho... 1
Helena & Frisco Mining Co., Gem, Idaho... 1
Riebold Mining Co., Grangeville, Idaho... 1
Wallace Electric Light Co., Wallace, Idaho... 3
Mullan Electric Lt. & Power Co., Mullan, Idaho 1
Milwaukee Mining Co., Wallace, Idaho, (displacing turbine) ... 1
Cœur d'Alene S. & L. Mining Co., Burke, Idaho 2
Cœur d'Alene M. & C. Co., Wallace, Idaho... 2
Ashland Electric Lt. & Power Co., Ashland, Or 2
Willis Saw Mill, Roseburg, Or. (displacing turbine) 1
Howell Stone Co., Roseburg, Or... 1
Columbia Packing Co., The Dalles, Or... 1
Wadleigh Drift Mine, Waldo, Or... 1
Elkhorn Mining Co., Baker Co., Or... 1
Senator Mine, Prescott, A. T... 1
U. S. Government, Fort Grant, A. T... 1
Jones' Mill, Prescott, A. T... 1
Apollo Con. Mining Co., Unga Island, Alaska... 3
Palo Alto Mining Co., Oroville, Cal... 1
Savannah Mine, Madera Co., California... 1
Los Angeles Water Wk's, Los Angeles, Cal. 1
Redlands El. L. & P. Co., Redlands, Cal... 7
Mt. Lowe Railway Co., Altadena, Cal ... 4
Weaverville El. Light Co., Weaverville, Cal. 1
Young Amer. Tailing Mill, Sierra City, Cal. 1
Golden River Mines, Placer Co., Cal... 1
Bonanza Mines Limited, Sierra City, Cal... 1
B. L. & A. Reduction Co., Union, Oregon... 1
Washington Red. Co., Coulee City, Wash... 1
Washington Br'k & L. Co., Spokane, Wash. 1
Blue Cañon Coal Co., New Whatcom, W... 1
Mexican Mining Co., Douglas Isl'd, Alaska 4
Alaska M. & M. Co., Douglas Isl'd, Alaska 4
Metlakahtla Ind. Co., Metlakahtla, Alaska.. 2
Juneau Min'g & Mfg. Co., Juneau, Alaska.. 1
Nowell Gold Mining Co., Juneau, Alaska... 3
Apollo Mining Co., Unga Island, Alaska... 1
Union Con. M. Co., Gibsonville, Cal... 1
Red Cross Lumber Co., Castle Crag, Cal... 2
Towle Bros., Lumber Mills, Towles, Cal... 6
Hickey Bros., Sawyer's Bar, California... 1
Rawhide Mine, Jamestown, California... 1
State Prison, Folsom, California... 1

## SOME UNITED STATES REFERENCES.

California Powder Works, Santa Cruz, Cal...40
Morrison & Jones Mine, Silverton, Colo..... 1
Milwaukee Mining Company, Gem, Idaho... 1
Azusa Ice Company, Los Angeles Co., Cal. 2
Bitter Root Devel'nt Co, Hamilton, Mont... 2
Joseph Johnson & Company, Arcata, Cal... 1
Fairhaven W. & P. Co., Fairhaven, Wash... 1
Western Beet Sugar Co., Watsonville, Cal.. 2
Lane & Hayward Mine, Juneau, Alaska...... 3
Tiger Mining Company, Burke, Idaho....... 1
Eureka & Excelsior Mine, Eureka, Oregon 2
Kelson Valley G. M. Co., Mojave, Cal....... 1
Leesburg G. M. & M. Co., Salmon C'y, Id... 1
Juneau Mining Company, Juneau, Alaska... 3
Golden Fleece M. & M. Co., Lake C'y, Colo. 1
Jackson & Lake View Mng. Co., Lundy, Cal. 1
Rumsey & Company, St. Louis, Missouri... 8
N. Almaden Qslvr. M. Co., N. Almad'n, Cal. 2
Daddy Gold Mining Company, Murray, Id. 1
Berners Bay M. & M. Co., Juneau, Alaska.. 4
American Dev. & Mining Co., Butte, Mont. 2
Salmon G. M. Company, Red Rock, A. T.. 1
Hoosier Mg. Company, Crested Butte, Colo. 1
Pioneer Mining Company, Colfax, Cal....... 1
Ogden Milling & Elevator Co., Ogden, Utah 1
Loaiza & Company, San Francisco, Cal...... 1
Chicago & Montana Min'g Co., Libby, Mont. 2
Harrison Mining Co., Silverton, Colo......... 1
Gallagher Mining Co., Johnsville, Cal......... 1
North Star Mining Co., Grass Valley, Cal... 2
Idaho and Yellow Jacket Mine, Idaho ........ 1
Esperanza Min'g Co., Mokelumne Hill, Cal. 4
United Verde Copper Co., Jerome, Arizona 2
Ophir Hill Mines, Salt Lake City, Utah...... 4
Tomichi Milling Company, Sargent, Colo.. 1
Atlanta G. & S. Con. Mines, Atlanta, Idaho 1
Gold Run Mining Company, Gilta, Cal....... 2
South Eureka Mining Co., Sutter Creek, Cal. 2
Idlewild G M. Co., El Dorado Co., Cal..... 3
Golden King M. Co., Alleghany C'y, Cal... 1
Empire M. &. M. Co., Empire, Colo.......... 1
Allen Gold Mining Co., La Plata, Colo...... 1
Bunker Hill & Sul. M. Co., Wardner, Idaho 3
Pioneer Quartz Mill Co., Newcastle, Cal..... 1
Monte Cristo Mining Co., Everett, Wash..... 1
Buena Vista P'ck'g Co., Buena Vista, Colo.. 1
Montana M. Co., Ltd., Marysville, Mont..... 2
Los Angeles Water Co., Los Angeles, Cal.. 1
Maryland Mining Company, Grass Val., Cal. 1
Occidental Con. M. Co., Virginia City, Nev. 1
Chrisman Bros., The Dalles, Oregon.... ..... 1
Carson Creek M. Co., Angels Camp, Cal... 1
Kinkhead M. & M. Co. Virginia City, Nev. 7
Alaska Mexican M. & M. Co., Juneau, Alas. 4
Columbian Fine G. Saving Co., Genesee, Id. 1
Champion Mining Co., Nevada Co., Cal..... 1
Pacific Steam Whaling Company, Alaska... 2
Union Con. G Mining Co., Pine Grove, Cal. 1
Central Eureka M. Co., Amador Co., Cal... 2
Towle Bros. Lumber Company, Towle, Cal. 2
Juneau Mining Company, Juneau, Alaska... 1
Jualin Gold Mining Co., Juneau, Alaska...... 2
Brown Bear Mining Co., Deadwood, Cal.... 1
Weston Mining Company, Grants Pass Or.. 1
Gwinn Mine Dev. Co., Calavaras Co., Cal... 5
Alaska M. & M. Co., Douglass Is., Alaska.. 1
Columbia G. & S. Mining Co., Yreka, Cal.. 1
Callow & Dall Mining Co., Silverton, Colo. 2
Highland Mining Co., Sierra Co., Cal........ 1
Garnet Gold Mining Co., Pony, Montana... 2
Bi-metallic Mining Co., Phillipsburg, Mont.. 2
Utah & Mont. M'ch'ry Co., S. L. C'y, Utah 2
Hector Mining Co., Mokelumne Hill, Cal... 1
Emerson G. M. Co., Mokelumne Hill, Cal.. 1
Heppner Flouring Mills, Heppner, Oregon 1
Metropolitan Mining Co., Nevada Co., Cal.. 2
El Encino Mining Co., Calavaras Co., Cal... 1
Mountain Mines, Limited, Keswick, Cal..... 1
Standard Mining Company, Wallace, Idaho 2
N. Bloomfield Mining Co., Nevada Co., Cal. 1
Allison Ranch Mine, Grass Valley, Cal....... 5
Oneida G. &. S. M. Co., Amador Co., Cal.. 1
Suffolk Globe M. & M. Co., Ophir, Colo.... 2
Perry Mining Company, Lathrop, Cal..... ... 2
Knapp, Titan Milling Co., Beatrice, Neb.... 1
Anaconda Copper M. Co., Hamilton, Mont. 1
Big Dipper Mining Co., Iowa Hill, Cal........ 1
Jamison Mining Company, Johnsville, Cal.. 3
White Cloud Gravel M. Co., Yreka, Cal..... 1
Keystone Mining Co., Yakh Dist., Mont...... 2
Bausch & Deming Co., Salmon City, Idaho. 1
Columbia Mining Company, Baker City, Or. 3
Butte Basin Gravel Mine, Jackson, Cal....... 1
Mayflower Gravel M. Co., Placer Co., Cal.. 1
Shasta Mining Company, Nevada Co., Cal.. 1
San Diego Water Company, San Diego, Cal. 1
Cadmus Mining Company, Nevada Co., Cal. 1
Ethel Mining Company, Magalia, Cal....... . 1
El Dorado Mining Co., El Dorado Co., Cal. 1
L. D. Ball Mine, Etna, California................ 1
Magalia Mine, Magalia, Butte Co., Cal........ 1
Roanoke Mining Co., Calaveras Co., Cal... 1
Quartz Glen Mine, Mokelumne Hill, Cal.... 2
Helena & Frisco Mining Co., Gem, Idaho... 5
P'tl'nd Alaska M. Co., Berner's Bay, Alaska 2
Diamond Mining Company, Placer Co., Cal. 1
Matterson & Williamson Co., Stockton, Cal. 1
Badger G. M. Co., Livingston, Idaho........... 2
Missoula Mercantile Co., Missoula, Mont.... 1
Black Oak M. Co., Tuolumne Co., Cal...... 1
Washington Mining Co., Nevada Co., Cal... 1
Nowell Gold Mining Co, Juneau, Alaska... 6
Helena Mining Co., Silver City, New Mexico 3
Four Hills Mining Co., Johnsville, Cal........ 5
Sierra Nevada W. & L. Co., Truckee, Cal.. 1
Ricks Water Company, Eureka, Cal........... 3
Gaston Ridge Mine, Graniteville, Cal......... 2
Rico Con. & Mining Company, Rico, Colo.. 1
Camp Bird Mines, Ouray, Colorado......... 1
Monte Cristo Mine, Trinity Co., Cal........... 1
American Improvement Co., Jackson, Cal... 2
Dexter Mining Company, Tuscarora, Nev... 1
Oneida Gold M. & M. Co., Amador Co., Cal.. 1
Melones Development Co., Melones, Cal.... 1
Republic Gold Mining Co., Spokane, Wash. 1
Lawrence Mining Company, Leland, Oregon 1
Gold Blossom Mining Co., Newcastle, Cal.. 1
Mastadon Mining Company, Pony, Montana 1
The Campagnie Francaise Hydraulic... .... 1
The Yreka Electric Light Co., Trinity Co... 1
Blymyer Iron Works, Cincinnati, Ohio...... 1
Crocker, Burbank & Co., Fitchburg, Mass.. 1
Philadelphia Engineering Works, Pa........ 1
Rix Compressed Air Machinery Co., S. F... 4
E. P. Allis Company, Milwaukee, Wis...... 9
Gates Iron Works, Chicago, Ill..................12
Union Iron Works, San Francisco, Cal......26
Fraser & Chalmers, Chicago, Ill................47

## SOME REFERENCES IN FOREIGN COUNTRIES.

Franco, Wilmanus & Co., Durango, Mex.. 1
Aurora Mine, Chiapas, Mexico.......... 1
Candelaria Mine, San Dimas, Mexico......... 7
Descubridora Mine, Cerro Verde, Mexico... 1
La Dura M. Co., Morelas, Chihuahua, Mex. 2
Tajo Mine, Rosario, Sinaloa, Mexico......... 1
Contra Estaca Min'g Co., Sinaloa, Mexico.. 1
Michoacan Ry. & Mg. Co., Michoacan, Mex. 1
Palmarejo Mining Company, Mexico......... 4
C. E. Mordaunt, Tapachula, Mexico......... 1
Santa Juliana Mine, Chihuahua, Mexico..... 1
A. Jacot, Temazcaltepec, Mexico.............. 2
Ventanas Con. Mining Company, Mexico... 3
Negociacion Minera de Jalisco, Mexico...... 1
San Antonio, Viejo M. Company, Mexico... 2
La Compañia Minera de Panuca, Mexico... 3
Ygnacio Amador, Topia, Durango, Mex..... 2
Paraiso Reduction Co., Honduras, C. A..... 3
New York and Rosario M. Co., Honduras.. 3
Olancho Mines, Limited, Honduras, C. A. 1
Central American Red. Co., Honduras......... 1
Oro Fino Mining Co., Honduras, C. A...... 1
Salto Mine, San Pedro, Honduras, C. A..... 1
Guayobillas M. Co., Yusearan, Honduras... 1
Guatemala Ice Co., Guatemala, C. A......... 1
Pellin Ice Works, Guatemala, C. A.. ........ 1
Fernando Conde Mine, Guatemala, C. A... 1
A. C. James, Churaj, Guatemala, C. A...... 1
Pablo Fournier, Ocos, Guatemala, C. A...... 1
La Trinidad Min'g Co., Costa Rica, C. A... 3
Costa Rica Min'g Co., Punta Arenas, C. A. 1
Finca La Vina Coffee Plant'n, Costa Rica... 1
Electric Light Co., San Jose dé Costa Rica 4
La Trinidad Mine, San Salvador, C. A...... 1
Mina La Union, San Salvador, C. A.......... 3
Campañia Luz Electrica, Santa Ana, San Salvador, C. A. (displacing turbines) 4
Santa Ana Water Works, San Salvador ..... 1
Minas de Libans, Tolima, U. S. Colombia.. 1
Bocaname Mine, Tolima, Colombia, S. A... 2
Guerra Hermanos. Ricaurte, Colombia, S.A. 1
El Christo Mng. Co., Tolima, Colombia, S.A. 2
Segovia M. Co., Medillon, Colombia, S. A. 2
Saviola Mining Co., Colombia, S. A. ........ 2
Pantaneo Mining Co., Colombia, S. A........ 1
Ospina Herms Medillon, Colombia, S. A... 1
Backus & Johnson, Lima, Peru, S. A...... 5
C. M. Pflucker & Bro., Lima, Peru......... 1
Casapalca Elec. Light Co., Casapalca, Peru 1
Thompson Houston Elec. Co., Pucara, Peru 1
Balfour & Co., Valparaiso, Chile, S. A...... 1
St. John del Rey Mining Co., Brazil, S. A... 7
Don Pedro Gold Mining Co., Brazil, S. A... 4
Petropolis El. Light Co., Brazil, S. A......... 4
San Vicente Mining Co., Mazatlan, Mex...... 1
Pinos Altos Mining Co., Chihuahua, Mex... 1
Colorado Iron Works, Guadaloupe, Mex... 1
G. Castaños, Guadalajara, Mexico........... 2
San Sebastian Mine, Mascota, Mexico....... 2
La Primavera Mine, Colcoman, Mexico...... 1
Candelaria Mex. Cons. Mining Co., Mex..... 2
Siamon Mines, Eracular, Amador, Mex..... 1
Negociacion Minar de Sianori, Sianori, Mex. 2
Batopilas Mining Co., Chihuahua, Mex...... 4
Woodward & Walker, Transvaal, S. Africa 1
Union Mine, Transvaal, South Africa......... 1
The Cerrilos Mining Co., Sinaloa, Mex...... 1
Hacienda de S. Domingo, Guatemala, C. A. 1
Mineral del Escondido, Chihuahua, Mexico 1

Del Monte M. & M. Co., Sonora, Mex........ 1
Pioneer Mine, Transvaal, South Africa...... 1
Nil Desperandum Mining Co., Transvaal... 1
Lysbon Berlyn Gold Field Co., S. Africa... 1
Mount Morgan Mine, Barberton, S. Africa... 1
Forbes Reef Gold Mg. Co., Transvaal, S. Af. 4
Moodie Mines, Transvaal, South Africa...... 2
Aepli & Gross, Santa Ana, Salvador, C. A. 1
Cia Alumbrado de Electrico de S. S., Salvador, C. A. (displacing turbines)..... 4
G. Boy, Mazatanango, Guatemala, C. A..... 1
Enrique Siegerist, Retalhuleu, Guatemala... 1
Segoria Mng. Co., Octal, Nicaragua, C. A... 4
Porto Cortez Mining Co, Honduras, C. A... 1
Segovia Mining Co, Nicaragua, C. A......... 1
Tinebi Light and Power Co., Costa Rica..... 2
Rosario Mining Co., Honduras, C. A......... 4
Cia Anomina Electrica de Bucaramanga, U. S. C., S. A. (displacing turbines) 4
Zancudo Mines, Antioquia, U. S. C., S. A... 2
Cia de Minas de Labano, Tolima, U. S. C... 1
Darien Gold M'ng Co., Cana., U. S. C., S. A. 2
Nieto & Co., Bogota, U. S. C., S. A..... ..... 1
Guarico Mining Co., Guarico, U. S. C., S. A. 1
C. M. Schroder & Co., Lima, Peru, S. A... 2
Mellendo Mining Co., Peru, S. A.............. 1
Carlos Mognoschigasuna, Cojatamba, Peru 1
Yavamina Mine, Arroyo Grande, Peru, S. A. 1
Pucara Mines, Cordilleras, Peru, S. A........ 1
Don Pedro Mine, Minas Geraes, Brazil, S. A. 5
St. John Del Rey Mining Co., Brazil, S. A... 1
Playa de Oro Mining Co., Ecuador, S. A... 1
Ferguson, Morgan & Co., Baranquila, S. A. 1
Hydraulic Engineering Co., Chester, Eng...37
North Wales, United Copper Mine, Eng..... 1
Spitzkop Min. Co., Ltd., London, England 1
Singrum Freres, Espinal, France.............. 1
Bear Valley Electric Co., Nova Scotia......... 1
Oldham Gold Co., Oldham, Nova Scotia... 1
Alex. Graham Bell, Cape Breton, N. S...... 1
Takata & Co., Yokohama, Japan.............. 2
Ashio Copper Mine, Tokyo, Japan....... ..... 1
Kioto-Fu-Cho Canal Works, Japan............ 5
Japanese Government, Kioto-Fu-Cho, Japan 2
S. Tenabe, city of Tokyo, Japan................ 2
T. Ishikawa, city of Kobe, Japan........... .. 1
Hawaiian Sugar Co., Honolulu, H. I......... 2
Honolulu Elec. Light Wks., Honolulu, H. I. 1
Laupahoehoe Sugar Company, H. I.......... 1
Union Iron Works, Honolulu, H. I............ 6
Spreckels & Bros.' Plantation, H. I........... 1
Phœnix Mine, New Zealand, Australia...... 2
Mandurama Gold M'g Co., Ltd., Australia.. 1
The Aropa Silver Mining Co., Australia..... 9
G. W. Bull, Auckland, New Zealand......... 1
Frederico Callares, Lisbon, Spain............. 1
Darjeeling Mines, Darjeeling, India........... 1
Burmah Ruby Mines, Burmah, India......... 3
J. Spencer Hollings, Montserrat, W. Indies 1
Geo. Armitage & Co., Colombo, Ceylon... 2
Sugar Mill, Tahiti, South Sea Islands ........ 1
City of Vancouver Pumping Station, B. C... 1
Wm. Crickmay, Vancouver, B. C......... .... 4
A. Laureston, Vancouver, B. C................. 1
Walter H. Kendall, Vancouver, B. C ........ 2
La Luz y San Juan Mining Co., Jalisco, Mex. 2
The Carrinto Mines, Michiogan, Mexico..... 1
Hacienda de S. Eulalia, Costa Rica, C. A. 1
Hacienda Aldama, Honduras, C. A... .... .. 1

## SOME REFERENCES IN FOREIGN COUNTRIES.

Aktieselskabet Cellulosefabrik, Norway..... 3
Heindal Maskinforretnung, Christiania, N'y 5
Saavedra, Benard & Co., Valparaiso, S. A. 1
Vargas y Hermanas, Colombia, S. A......... 2
G. & O. Braniff & Co., C'y of Mexico, M... 1
Foulds Freres, Bogota, Colombia, S. A...... 1
Alberto Moreno, San Jose, Costa Rica, C. A. 1
Carlos J. Canal, Cucuta, Colombia, S. A..... 3
Jose A. Borges, Caraboro, Venezuela, S. A. 1
Uricochea & Gomez, Colombia, S. A......... 1
Ramon Baez, hijo., Caracas, Venez., S. A.. 4
Sr. Don Luis Nieto, Colombia, S. A.......... 1
Alejandro Cacaerez, Colombia, S. A......... 1
Walter S. Riote, Costa Rica, C. A........... 1
Frederico Ortez, Punta Arenas, C. A ........ 1
A. Parado, San Antonio, Mexico............... 1
Miguel Nieto, Ricaurte, Colombia, S. A..... 1
Manuel Jose Abandano, Bogota, Col., S. A. 1
Ditta Allessandro Calzoni, Bologna, Italy... 1
S. A. Development Co., Guayaquil, S. A... 3
Atlas Mine, Greytown, Nicaragua, C. A..... 1
Alberto G. Belderraine, Oaxaca, Mexico...... 1
Rand Drill Company, Sydney, Australia..... 1
Jose Gomez & Co., Bogota, Colombia, S. A. 1
Alejandro Urdaneta, Bogota, Col., S. A..... 1
Juan Bannister, City of Mexico, Mexico...... 1
Antonio Zolli & Company, Milan, Italy...... 6
Jose J. Aranjo, Bogota, Colombia, S. A...... 1
Rafael Roca, Girardot, South America....... 1
Herrerria y Sta Maria, Girardot, S. A......... 1
Marcus Mason & Company, Central America 3
Glynnburg Mine, Transvaal, South Africa... 1
S. P. Nicolson & Co., Rio Janeiro, S. A..... 2
Hidalgo Mining Company, Hildago, Mexico 1
Sydney Power Company, Sydney, N. S. W. 3
Lanzagorta & Company, San Blas, Mexico.. 2
Campbell & Anderson, N. Westmstr., B. C. 1
Hacienda Mojarres, San Blas, Mexico........ 1
Atlanta Vista Coffee Plantation, Guatemala 1
Frco. Gemoz Ortizoga, Puerto, Angel, C. A. 1
Furbach, Dietze & Co., Tapachula, Mexico 1
Compania Union Occidental, Acajutla, Sal. 1
Ortiga & Scholz Hacienda, Tapachula, M... 1
De Laguna Coffee Plantation, Ocos, Mexico 1
J. M. Paras Ballestros, Montemorlos, Mexico 1
Delius & Company, Tepic, Mexico........... 1
Camille Licardie, San Marco, Guatemala... 2
Tuman ler & Weibalch, S. J. de Guatemala 1
Slocan Star Mining Co., Three Forks, B. C. 1
C. Pacific M. &. M. Co., Ainsworth, B. C. 2
Takata & Company, Tokio, Japan............. 3
Rossland L., W. & P. Co., Rossland, B. C. 6
Latonia, Imboden & Co., Acajutla, Salvador 1
Minera de Santiago, Tultenango, Mexico.... 1
Bolanos Hermanos, S. J. de Guatemala, C.A. 1
John T. Wright, Guatemala, C. A............. 1
Herman Wundrum, Ocos, Guatemala, C. A. 1
Finca el Quezal, Champerico, C. A.......... 1
A. J. Northcraft, Jimenez, Mexico.......... 1
La Florida Coffee Plantation, Guat., C. A... 2
La Lux Coffee Plantation, Guat., C. A...... 1
Don Carlos d'Aubisson, San Salvador, C. A. 1
Cia del Ferrocarril, Angangueo, Mexico..... 1
Hakalau Plantation Company, Hawaii, H. I. 1
Station Cruz de Yojos, Spanish Hon., C. A. 1
P. L. Barrenguy, Concordia, Oaxaca, Mexico 2
Santa Edwiges M. Co., Chihuahua, Mexico 1
Hilo Sugar Company, Hilo, Hawaii........... 1
Zadik & Cheeseman, Quezaltenango, Mex.. 2
J. R. Partridge, Tepic, Sonora, Mexico...... 3

Noble Five Con. M. &. M. Co., Cody, B. C. 2
Morehouse & Morrill, Quezaltenango, Guat. 2
Goedang Tjihap Coffee Ftry., Tjibadak, Java 1
Pasir Karet Coffee Factory, Tjibadak, Java 1
Hambalang Coffee Factory, Sinagar, Java... 1
Knox, Schlapp & Co., Melbourne, Australia 1
Guillermo Fisher, Tuxtla, Mexico............. 1
Tilanros Coffee Company, Corinto, Mexico 1
Kootenaian Mining Company, Kaslo, B. C. 1
Pinos Altos Mng. Co., Pinos Altos, Mexico 1
Pedrick & Ayre, Topia, Durango, Mexico.. 1
Don Leon Alvila, San Salvador, C. A........ 1
Enterprise Mining Co., Roseberry, B. C...... 1
Herbert Smith & Co., Guatemala, C. A..... 1
Rudulfo Grunberger, Mexico, Mexico....... 1
Galena Gold Mining Co., Silverton, B. C.... 2
Don Gus Sempe, S. J. de Guatemala, C. A. 1
Guido Richter & Co., Oaxaca, Mexico........ 1
Cen. Am. Government, Acajutla, C. A....... 1
Armstrong & Morrison, Vancouver, B. C...... 1
Maximo Weinburg, Quezaltenango, C. A... 1
La Cumbre G. M. Co., Chihuahua, Mexico.. 2
Malabar Tea Estate, Sinagar, Tjibadak, Java 1
Thos. Bateman & Son, Launceston, Tas..... 2
Larco Brothers, Frujillo, Peru, S. A............ 1
E. Basadre Forero, La Paz, Bolivia, S. A... 1
Compania Indus. Mex'o, Chihuahua, Mexico 1
San Vicente Mining Co., Sinaloa, Mexico.... 1
Anchor Tin Mines, Tasmania, Australia .. .. 2
New Zealand Crown Mines, Ltd., Australia 4
Le Roi Mng. & Smelting Co., Rossland, B. C. 1
Kaslo M. & M. Co., South Forks, B. C....... 1
Byron White Company, Sandon, B. C........ 1
Frank S. Barnard, Vancouver, B. C........... 1
Nelson Electric Light Co., Nelson, B. C..... 1
Sandon Water Works Co., Sandon, B. C.... 2
Domingo Mines, Limited, Brazil, S. A....... 2
Minas de Plata Y Cobre, Chile, S. A......... 1
Potosi Mines, Limited, Bolivia, S. A......... 1
Esmeralda Mining Co., Ecuador, S. A....... 1
Hilo Ice Works, Hilo, Hawaiian Islands..... 1
Mercede Mining Company, U. S. Colombia 1
Conception Mining Co., Ltd., Paraguay...... 1
Huamagua Mines, Limited, Peru, S. A...... 2
Waterous Engine Works, Brantford, Canada 2
Plantagenet Mines, Ontario, Canada.......... 1
Wellington Mines, Ltd., Ontario, Canada.. 1
Bengal Mft. Company, British India........ .. 1
Allendale Mining Co., Ontario, Canada...... 1
San Rafael M. Co., Ltd., Venezuela, S. A.. 1
Samuel H. Fisk, Monterey, Mexico........... 3
Backus & Johnson, Lima, Peru, S. A......... 6
Parke & Lacy Company, Sydney, Australia 9
Spencer & Waters, Santiago, Chile, S. A... 6
Bagnall & Hilles, Yokohama, Japan ..........12
G. A. Ritchie, Bogota, U. S. Colombia.......16
Manning Brothers, Matagalpa, Nicaragua... 2
A. R. W. Kerkhoven, island of Java.........12
Malingoet Tea Plantation, island of Java..... 1
San Lorenzo Mines, Ltd., Paraguay, S. A... 1
Panaco Mining Co., Yucatan, Mexico....... 1
Campamento Mines, Honduras, C. A......... 1
Hacienda Olancho, Honduras, C. A.......... 1
Chimbarazo Mines, Ecuador, S. A............. 1
Santa Cruz Mining Company, Brazil, S A.. 1
Cordillera Grande Mines, Brazil, S. A........ 1
Enrique Seigerist, Retalhuleu, Guatemala... 8
Grace & Company, Valparaiso, Chile.......... 5
Nieto & Co., United States Colombia, S. A. 1
Fraser & Chalmers, London, England...... .30

# THE McCORMICK REGISTER GATE TURBINE.

The print herewith shows a pair of 50-inch McCormick Turbines of 1,300 horse power capacity, running under 26 feet head, with a sheave 10 feet in diameter, carrying twenty-seven 1½-inch ropes. The power supplied from these wheels operates a General Electric Co.'s generator at number 2 Station of the Sacramento Gas, Electric and Railway Co., at Folsom, Cal.

This Company is also running four pair of 30-inch horizontal Turbines of same make; aggregating 5,000 horse-power. These wheels operate under 55 feet head and are direct connected to General Electric Co.'s generators—three phase type—of corresponding capacity.

The current from this station—aggregating some 6,000 horse-power—is transmitted to Sacramento, 22 miles distant, at pressure of 11,000 volts, and used for running an extensive railroad system and for general manufacturing purposes.

This is regarded as one of the most successful transmission plants so far installed and the largest in this country with the single exception of that at Niagara Falls.

This wheel is offered for the purpose of utilizing lower heads than the PELTON can be operated under to advantage. The uniformly good results obtained from the McCormick Turbines in every variety of service and under widely varying conditions give assurance that they embrace the best features of turbine practice, and that the construction is such as to afford all possible security against irregular and unreliable work.

The advantages claimed for this wheel are briefly stated—simplicity of construction, high efficiency, both at full and part gate, unusual strength, and consequent durability. These wheels are guaranteed to develop eighty per cent efficiency under all conditions of service, while in the majority of cases they will run somewhat higher.

Governors are furnished which are guaranteed to afford close and sensitive control under all variations of load.

## PRICE LIST OF PULLEYS.

WROUGHT RIM AND WOOD SPLIT PULLEYS.

| Diameter in inches | Width of face in inches: 3 | 4 | 5 | 6 | 7 | 8 | 9 | 10 | 11 | 12 | 13 | 14 | 15 | 16 | 17 | 18 | 19 | 20 | 21 | 22 | 23 | 24 | 25 |
|---|---|---|---|---|---|---|---|---|---|---|---|---|---|---|---|---|---|---|---|---|---|---|---|
| 3 | 1 90 | 2 00 | 2 10 | 2 25 | 2 40 | 2 60 | 2 85 | 3 10 | 3 35 | 3 60 | 3 85 | 4 10 | 4 40 | 4 75 | | | | | | | | | |
| 4 | 2 00 | 2 10 | 2 20 | 2 40 | 2 60 | 2 85 | 3 10 | 3 35 | 3 60 | 3 85 | 4 10 | 4 40 | 4 75 | 5 10 | | | | | | | | | |
| 5 | 2 10 | 2 20 | 2 40 | 2 60 | 2 85 | 3 10 | 3 35 | 3 60 | 3 85 | 4 10 | 4 40 | 4 75 | 5 10 | 5 45 | | | | | | | | | |
| 6 | 2 20 | 2 35 | 2 55 | 2 75 | 3 00 | 3 25 | 3 50 | 3 75 | 4 00 | 4 30 | 4 65 | 5 00 | 5 35 | 5 70 | | | | | | | | | |
| 7 | 2 30 | 2 45 | 2 65 | 2 85 | 3 10 | 3 35 | 3 60 | 3 85 | 4 15 | 4 50 | 4 85 | 5 20 | 5 55 | 5 90 | | | | | | | | | |
| 8 | 2 45 | 2 60 | 2 80 | 3 00 | 3 25 | 3 50 | 3 75 | 4 00 | 4 30 | 4 65 | 5 00 | 5 40 | 5 80 | 6 20 | | | | | | | | | |
| 9 | 2 60 | 2 80 | 3 00 | 3 25 | 3 50 | 3 75 | 4 00 | 4 30 | 4 65 | 5 00 | 5 40 | 5 80 | 6 20 | 6 60 | 7 00 | 7 50 | 8 00 | 8 50 | | | | | |
| 10 | 2 80 | 3 00 | 3 25 | 3 50 | 3 75 | 4 00 | 4 30 | 4 65 | 5 00 | 5 40 | 5 80 | 6 20 | 6 60 | 7 00 | 7 50 | 8 00 | 8 60 | 9 20 | | | | | |
| 11 | 3 00 | 3 25 | 3 50 | 3 75 | 4 00 | 4 30 | 4 65 | 5 00 | 5 40 | 5 80 | 6 20 | 6 60 | 7 00 | 7 50 | 8 00 | 8 60 | 9 20 | 9 80 | | | | | |
| 12 | 3 15 | 3 40 | 3 65 | 3 90 | 4 20 | 4 55 | 4 90 | 5 30 | 5 70 | 6 10 | 6 50 | 6 90 | 7 40 | 7 90 | 8 50 | 9 10 | 9 70 | 10 40 | 11 10 | 11 80 | 12 60 | 13 40 | |
| 13 | 3 30 | 3 55 | 3 80 | 4 10 | 4 45 | 4 80 | 5 20 | 5 60 | 6 00 | 6 40 | 6 80 | 7 30 | 7 80 | 8 40 | 9 00 | 9 60 | 10 30 | 11 00 | 11 70 | 12 50 | 13 30 | 14 10 | |
| 14 | 3 45 | 3 70 | 4 00 | 4 35 | 4 70 | 5 10 | 5 55 | 6 00 | 6 45 | 6 90 | 7 40 | 7 90 | 8 50 | 9 10 | 9 70 | 10 40 | 11 10 | 11 80 | 12 60 | 13 40 | 14 20 | 15 00 | |
| 15 | 3 60 | 3 90 | 4 25 | 4 60 | 5 00 | 5 45 | 5 90 | 6 35 | 6 80 | 7 30 | 7 80 | 8 40 | 9 00 | 9 60 | 10 30 | 11 00 | 11 80 | 12 60 | 13 40 | 14 20 | 15 10 | 16 00 | 16 90 |
| 16 | 3 80 | 4 15 | 4 50 | 5 00 | 5 50 | 6 00 | 6 50 | 7 00 | 7 50 | 8 00 | 8 50 | 9 00 | 9 60 | 10 30 | 11 00 | 11 80 | 12 60 | 13 40 | 14 20 | 15 10 | 16 00 | 16 90 | 17 80 |
| 17 | 4 00 | 4 40 | 4 90 | 5 40 | 5 90 | 6 40 | 6 90 | 7 40 | 7 90 | 8 40 | 8 90 | 9 50 | 10 20 | 11 00 | 11 80 | 12 60 | 13 40 | 14 20 | 15 10 | 16 00 | 16 90 | 17 80 | 18 80 |
| 18 | 4 20 | 4 70 | 5 20 | 5 70 | 6 20 | 6 70 | 7 20 | 7 80 | 8 40 | 9 00 | 9 60 | 10 30 | 11 10 | 11 90 | 12 70 | 13 50 | 14 30 | 15 20 | 16 10 | 17 00 | 17 90 | 19 00 | 20 10 |
| 19 | 4 40 | 4 90 | 5 40 | 5 90 | 6 50 | 7 10 | 7 70 | 8 30 | 8 90 | 9 50 | 10 20 | 11 00 | 11 80 | 12 60 | 13 40 | 14 30 | 15 20 | 16 10 | 17 00 | 17 90 | 19 00 | 20 10 | 21 20 |
| 20 | 4 60 | 5 20 | 5 80 | 6 40 | 7 00 | 7 60 | 8 30 | 9 00 | 9 70 | 10 40 | 11 20 | 12 00 | 12 90 | 13 80 | 14 70 | 15 60 | 16 50 | 17 50 | 18 50 | 19 60 | 20 70 | 21 80 | 22 90 |
| 21 | 4 80 | 5 40 | 6 00 | 6 60 | 7 30 | 8 00 | 8 70 | 9 40 | 10 20 | 11 00 | 11 80 | 12 70 | 13 60 | 14 50 | 15 50 | 16 50 | 17 50 | 18 60 | 19 70 | 20 80 | 22 00 | 23 20 | 24 40 |
| 22 | 5 00 | 5 60 | 6 30 | 7 00 | 7 70 | 8 40 | 9 20 | 10 00 | 10 80 | 11 70 | 12 60 | 13 50 | 14 50 | 15 50 | 16 60 | 17 70 | 18 80 | 20 00 | 21 20 | 22 40 | 23 60 | 24 80 | 26 00 |
| 23 | 5 20 | 5 80 | 6 50 | 7 20 | 8 00 | 8 80 | 9 70 | 10 60 | 11 50 | 12 50 | 13 60 | 14 70 | 15 80 | 17 00 | 18 20 | 19 40 | 20 60 | 21 80 | 23 00 | 24 30 | 25 60 | 26 90 | 28 20 |
| 24 | 5 50 | 6 20 | 7 00 | 7 80 | 8 60 | 9 40 | 10 30 | 11 20 | 12 10 | 13 00 | 14 10 | 15 30 | 16 50 | 17 70 | 18 90 | 20 20 | 21 50 | 22 80 | 24 10 | 25 40 | 26 80 | 28 20 | 29 60 |
| 25 | 5 90 | 6 60 | 7 30 | 8 10 | 9 00 | 10 00 | 11 00 | 12 00 | 13 00 | 14 00 | 15 20 | 16 40 | 17 70 | 19 00 | 20 40 | 21 80 | 23 20 | 24 60 | 26 00 | 27 50 | 29 00 | 30 50 | 32 00 |
| 26 | 6 20 | 7 00 | 7 80 | 8 60 | 9 50 | 10 50 | 11 50 | 12 60 | 13 75 | 15 00 | 16 30 | 17 60 | 19 00 | 20 40 | 21 80 | 23 30 | 24 80 | 26 30 | 27 90 | 29 50 | 31 10 | 32 70 | 34 30 |
| 27 | 6 60 | 7 40 | 8 20 | 9 00 | 10 00 | 11 10 | 12 20 | 13 40 | 14 60 | 15 80 | 17 10 | 18 50 | 20 00 | 21 50 | 23 00 | 24 50 | 26 10 | 27 70 | 29 30 | 30 90 | 32 50 | 34 10 | 35 70 |
| 28 | 7 00 | 7 80 | 8 60 | 9 50 | 10 50 | 11 60 | 12 80 | 14 00 | 15 20 | 16 50 | 17 90 | 19 40 | 20 90 | 22 40 | 24 00 | 25 60 | 27 20 | 28 80 | 30 40 | 32 00 | 33 60 | 35 30 | 37 00 |
| 29 | 7 30 | 8 20 | 9 10 | 10 00 | 11 10 | 12 30 | 13 50 | 14 75 | 16 00 | 17 50 | 19 00 | 20 50 | 22 10 | 23 70 | 25 30 | 27 00 | 28 70 | 30 40 | 32 10 | 33 80 | 35 50 | 37 20 | 38 90 |
| 30 | 7 70 | 8 60 | 9 50 | 10 60 | 11 80 | 13 00 | 14 30 | 15 60 | 17 00 | 18 50 | 20 00 | 21 50 | 23 25 | 25 00 | 26 75 | 28 50 | 30 25 | 32 00 | 33 75 | 35 50 | 37 25 | 39 00 | 40 75 |

Pulleys of wider face than 18 inches are provided with two sets arms without extra charge. When Pulleys of 18-inch face and under are ordered with two sets arms, an extra charge is made; a single set of arms for such widths, when made of our heavy patterns, is amply strong for any ordinary strain. We supply each Pulley with two set screws without charge. A moderate charge is made for extra set screws, key seats, babbitting or for extra large bore in proportion to the size of Pulley.

# PRICE LIST OF PULLEYS.

WIDTH OF FACE IN INCHES.

| DIAMETER IN INCHES. | 3 | 4 | 5 | 6 | 7 | 8 | 9 | 10 | 11 | 12 | 13 | 14 | 15 | 16 | 17 | 18 | 19 | 20 | 21 | 22 | 23 | 24 | 25 | DIAMETER IN INCHES. |
|---|---|---|---|---|---|---|---|---|---|---|---|---|---|---|---|---|---|---|---|---|---|---|---|---|
| 31 | 8 00 | 9 00 | 10 00 | 11 10 | 12 35 | 13 60 | 15 00 | 16 50 | 18 00 | 19 60 | 21 20 | 22 80 | 24 60 | 26 40 | 28 20 | 30 00 | 31 80 | 33 60 | 35 40 | 37 20 | 39 00 | 40 80 | 42 60 | 31 |
| 32 | 8 50 | 9 50 | 10 50 | 11 60 | 12 85 | 14 10 | 15 50 | 17 00 | 18 60 | 20 30 | 22 00 | 23 80 | 25 60 | 27 40 | 29 20 | 31 00 | 33 00 | 35 00 | 37 00 | 39 00 | 41 00 | 43 00 | 45 00 | 32 |
| 33 | 9 00 | 10 00 | 11 15 | 12 35 | 13 65 | 15 00 | 16 45 | 18 00 | 19 65 | 21 40 | 23 10 | 24 95 | 26 75 | 28 60 | 30 40 | 32 30 | 34 35 | 36 40 | 38 50 | 40 60 | 42 70 | 44 90 | 47 10 | 33 |
| 34 | 9 40 | 10 60 | 11 80 | 13 10 | 14 45 | 15 90 | 17 40 | 19 00 | 20 70 | 22 50 | 24 20 | 26 10 | 27 90 | 29 80 | 31 70 | 33 60 | 35 70 | 37 80 | 40 00 | 42 20 | 44 40 | 46 80 | 49 20 | 34 |
| 35 | 9 85 | 11 15 | 12 45 | 13 85 | 15 25 | 16 80 | 18 35 | 20 00 | 21 70 | 23 50 | 25 30 | 27 20 | 29 10 | 31 00 | 32 90 | 34 90 | 37 00 | 39 20 | 41 50 | 43 80 | 46 10 | 48 70 | 51 30 | 35 |
| 36 | 10 30 | 11 70 | 13 10 | 14 60 | 16 10 | 17 70 | 19 30 | 21 00 | 22 00 | 24 00 | 26 00 | 28 00 | 30 00 | 32 00 | 34 00 | 36 00 | 38 00 | 40 00 | 43 00 | 45 00 | 47 00 | 50 00 | 53 00 | 36 |
| 38 | | 12 80 | 14 40 | 16 00 | 17 70 | 19 50 | 21 20 | 23 00 | 24 50 | 26 00 | 28 00 | 30 00 | 32 00 | 34 00 | 36 00 | 38 00 | 41 00 | 43 00 | 46 00 | 48 00 | 51 00 | 54 00 | 57 00 | 38 |
| 40 | | 13 90 | 15 70 | 17 50 | 19 30 | 21 25 | 23 10 | 25 00 | 26 50 | 28 00 | 30 00 | 32 00 | 34 00 | 36 00 | 39 00 | 41 00 | 43 00 | 46 00 | 49 00 | 51 00 | 54 00 | 58 00 | 61 00 | 40 |
| 42 | | 15 00 | 17 00 | 19 00 | 21 00 | 23 00 | 25 00 | 27 00 | 29 00 | 31 00 | 33 00 | 35 00 | 37 00 | 39 00 | 41 00 | 44 00 | 46 00 | 49 00 | 52 00 | 55 00 | 58 00 | 62 00 | 66 00 | 42 |
| 44 | | 17 00 | 19 00 | 21 00 | 23 00 | 25 00 | 27 00 | 29 00 | 31 00 | 33 00 | 35 00 | 37 00 | 40 00 | 42 00 | 44 00 | 47 00 | 50 00 | 53 00 | 56 00 | 60 00 | 63 00 | 67 00 | 71 00 | 44 |
| 46 | | 19 00 | 21 00 | 23 00 | 25 00 | 27 00 | 29 00 | 31 00 | 33 00 | 36 00 | 38 00 | 40 00 | 43 00 | 45 00 | 48 00 | 51 00 | 54 00 | 57 00 | 61 00 | 65 00 | 68 00 | 73 00 | 77 00 | 46 |
| 48 | | 21 00 | 23 00 | 25 00 | 27 00 | 29 00 | 31 50 | 33 90 | 36 20 | 38 50 | 40 00 | 43 00 | 46 00 | 48 00 | 51 00 | 54 00 | 58 00 | 61 00 | 65 00 | 70 00 | 74 00 | 78 00 | 83 00 | 48 |
| 50 | | | 25 00 | 27 00 | 29 00 | 31 00 | 33 70 | 36 20 | 38 60 | 41 00 | 43 00 | 46 00 | 49 00 | 51 00 | 54 00 | 58 00 | 62 00 | 65 00 | 70 00 | 75 00 | 79 00 | 84 00 | 89 00 | 50 |
| 52 | | | 28 00 | 30 00 | 32 00 | 34 00 | 36 00 | 38 50 | 41 00 | 43 50 | 46 00 | 49 00 | 52 00 | 55 00 | 58 00 | 62 00 | 66 00 | 70 00 | 75 00 | 80 00 | 85 00 | 90 00 | 95 00 | 52 |
| 54 | | | 30 00 | 32 00 | 34 00 | 36 00 | 38 50 | 41 20 | 43 80 | 46 00 | 49 00 | 52 00 | 55 00 | 59 00 | 62 00 | 66 00 | 70 00 | 75 00 | 80 00 | 85 00 | 90 00 | 95 00 | 100 00 | 54 |
| 56 | | | | 35 00 | 37 00 | 39 00 | 41 00 | 43 00 | 46 00 | 49 00 | 52 00 | 55 00 | 59 00 | 63 00 | 66 00 | 71 00 | 75 00 | 80 00 | 85 00 | 90 00 | 95 00 | 101 00 | 106 00 | 56 |
| 58 | | | | 38 00 | 40 00 | 42 00 | 44 00 | 46 00 | 49 00 | 52 00 | 55 00 | 59 00 | 62 00 | 67 00 | 71 00 | 75 00 | 80 00 | 85 00 | 90 00 | 95 00 | 101 00 | 106 00 | 112 00 | 58 |
| 60 | | | | 41 00 | 43 00 | 45 00 | 47 00 | 49 00 | 52 00 | 55 00 | 58 00 | 62 00 | 66 00 | 71 00 | 75 00 | 80 00 | 85 00 | 90 00 | 95 00 | 100 00 | 106 00 | 112 00 | 118 00 | 60 |
| 62 | | | | 44 00 | 46 00 | 48 00 | 50 00 | 52 00 | 55 00 | 58 00 | 62 00 | 66 00 | 70 00 | 75 00 | 80 00 | 85 00 | 90 00 | 95 00 | 100 00 | 106 00 | 112 00 | 118 00 | 124 00 | 62 |
| 64 | | | | 46 00 | 48 00 | 50 00 | 52 00 | 55 00 | 58 00 | 61 00 | 65 00 | 69 00 | 74 00 | 79 00 | 84 00 | 89 00 | 94 00 | 99 00 | 104 00 | 110 00 | 117 00 | 123 00 | 130 00 | 64 |
| 66 | | | | 48 00 | 50 00 | 52 00 | 54 00 | 57 00 | 61 00 | 64 00 | 69 00 | 73 00 | 78 00 | 83 00 | 88 00 | 93 00 | 98 00 | 103 00 | 109 00 | 115 00 | 122 00 | 129 00 | 136 00 | 66 |
| 68 | | | | 50 00 | 52 00 | 54 00 | 56 00 | 60 00 | 64 00 | 68 00 | 72 00 | 77 00 | 82 00 | 87 00 | 92 00 | 97 00 | 102 00 | 108 00 | 113 00 | 120 00 | 127 00 | 134 00 | 142 00 | 68 |
| 70 | | | | 52 00 | 54 00 | 56 00 | 59 00 | 63 00 | 67 00 | 71 00 | 76 00 | 81 00 | 86 00 | 91 00 | 96 00 | 101 00 | 106 00 | 112 00 | 118 00 | 125 00 | 132 00 | 140 00 | 148 00 | 70 |
| 72 | | | | 54 00 | 56 00 | 58 00 | 62 00 | 66 00 | 70 00 | 75 00 | 80 00 | 85 00 | 90 00 | 95 00 | 100 00 | 105 00 | 111 00 | 117 00 | 123 00 | 130 00 | 138 00 | 146 00 | 155 00 | 72 |
| 78 | | | | | | 71 00 | 75 00 | 80 00 | 85 00 | 90 00 | 95 00 | 100 00 | 105 00 | 110 00 | 116 00 | 121 00 | 127 00 | 133 00 | 140 00 | 147 00 | 155 00 | 164 00 | 173 00 | 78 |
| 84 | | | | | | 85 00 | 90 00 | 96 00 | 101 00 | 107 00 | 112 00 | 118 00 | 123 00 | 130 00 | 136 00 | 143 00 | 149 00 | 156 00 | 163 00 | 170 00 | 179 00 | 188 00 | 197 00 | 84 |
| 90 | | | | | | 100 00 | 107 00 | 114 00 | 121 00 | 128 00 | 135 00 | 142 00 | 149 00 | 157 00 | 165 00 | 173 00 | 181 00 | 189 00 | 197 00 | 206 00 | 215 00 | 225 00 | 235 00 | 90 |
| 96 | | | | | | 115 00 | 123 00 | 132 00 | 140 00 | 149 00 | 157 00 | 166 00 | 174 00 | 184 00 | 193 00 | 203 00 | 212 00 | 222 00 | 231 00 | 241 00 | 251 00 | 262 00 | 272 00 | 96 |
| | 3 | 4 | 5 | 6 | 7 | 8 | 9 | 10 | 11 | 12 | 13 | 14 | 15 | 16 | 17 | 18 | 19 | 20 | 21 | 22 | 23 | 24 | 25 | |

INSTRUCTIONS FOR ORDERING.—Give diameter, width of face, size of bore, and state whether crowning or flat face. Tight and Loose Pulleys should be crowning and the one driving them should be flat face and as wide as both. We also make SPLIT PULLEYS, for which add 20 per cent to this price list. Tight and Loose Pulleys have long hubs of equal length, both tapped for set-screws, and are interchangeable when worn on shaft. For these add 10 per cent to this list.

# POWER TRANSMITTING TABLES.

## Transmitting Efficiency of Turned Iron Shafting at Different Speeds.

### As Prime Mover or Head Shaft carrying Main Driving Pulley or Gear, well supported by bearings.

| Diameter of Shaft. | Number of Revolutions per Minute. | | | | | | | | | | |
|---|---|---|---|---|---|---|---|---|---|---|---|
| | 60 | 80 | 100 | 125 | 150 | 175 | 200 | 225 | 250 | 275 | 300 |
| Inches. | H.P. | H.P. | H.P. | H.P. | H.P. | H.P. | H.P. | H.P. | H.P. | H.P. | H.P. |
| 1¾ | 2.6 | 3.4 | 4.3 | 5.4 | 6.4 | 7.5 | 8.6 | 9.7 | 10.7 | 11.8 | 12.9 |
| 2 | 3.8 | 5.1 | 6.4 | 8 | 9.6 | 11.2 | 12.8 | 14.4 | 16 | 17.6 | 19.2 |
| 2¼ | 5.4 | 7.3 | 8.1 | 10 | 12 | 14 | 16 | 18 | 20 | 22 | 24 |
| 2½ | 7.5 | 10 | 12.5 | 15 | 18 | 22 | 25 | 28 | 31 | 34 | 37 |
| 2¾ | 10 | 13 | 16 | 20 | 24 | 28 | 32 | 36 | 40 | 44 | 48 |
| 3 | 13 | 17 | 20 | 25 | 30 | 35 | 40 | 45 | 50 | 55 | 60 |
| 3¼ | 16 | 22 | 27 | 34 | 40 | 47 | 54 | 61 | 67 | 74 | 81 |
| 3½ | 20 | 27 | 34 | 42 | 51 | 59 | 68 | 76 | 85 | 94 | 102 |
| 3¾ | 25 | 33 | 42 | 52 | 63 | 73 | 84 | 94 | 105 | 115 | 126 |
| 4 | 30 | 41 | 51 | 64 | 76 | 89 | 102 | 115 | 127 | 140 | 153 |
| 4½ | 43 | 58 | 72 | 90 | 108 | 126 | 144 | 162 | 180 | 198 | 216 |
| 5 | 60 | 80 | 100 | 125 | 150 | 75 | 200 | 225 | 250 | 275 | 300 |
| 5½ | 80 | 106 | 133 | 166 | 199 | 233 | 266 | 299 | 333 | 266 | 400 |

### As Second Movers or Line Shafting. Bearings 8 feet apart.

| Diameter of Shaft. | Number of Revolutions per Minute. | | | | | | | | | | |
|---|---|---|---|---|---|---|---|---|---|---|---|
| | 100 | 125 | 150 | 175 | 200 | 225 | 250 | 275 | 300 | 325 | 350 |
| Inches. | H.P. | H.P. | H.P. | H.P. | H.P. | H.P. | H.P. | H.P. | H.P. | H.P. | H.P. |
| 1¾ | 6 | 7.4 | 8.9 | 10.4 | 11.9 | 13.4 | 14.9 | 16.4 | 17.9 | 19.4 | 20.9 |
| 1⅞ | 7.3 | 9.1 | 10.9 | 12.7 | 14.5 | 16.3 | 18.2 | 20.0 | 21.8 | 23.6 | 25.4 |
| 2 | 8.9 | 11.1 | 13.3 | 15.5 | 17.7 | 20.0 | 22.2 | 24.4 | 26.6 | 28.8 | 31 |
| 2⅛ | 10.6 | 13.2 | 15.9 | 18.5 | 21.2 | 23.8 | 26.5 | 29.1 | 31.8 | 34.4 | 37 |
| 2¼ | 12.6 | 15.8 | 19 | 22 | 25 | 28 | 31 | 35 | 38 | 41 | 44 |
| 2⅜ | 15 | 18 | 22 | 26 | 29 | 33 | 37 | 41 | 44 | 48 | 52 |
| 2½ | 17 | 21 | 26 | 30 | 34 | 39 | 43 | 47 | 52 | 56 | 60 |
| 2¾ | 23 | 29 | 34 | 40 | 46 | 52 | 58 | 64 | 69 | 75 | 81 |
| 3 | 30 | 37 | 45 | 52 | 60 | 67 | 75 | 82 | 90 | 97 | 105 |
| 3¼ | 38 | 47 | 57 | 66 | 76 | 85 | 95 | 104 | 114 | 123 | 133 |
| 3½ | 47 | 59 | 71 | 83 | 95 | 107 | 119 | 131 | 143 | 155 | 167 |
| 3¾ | 58 | 73 | 88 | 102 | 117 | 132 | 146 | 162 | 176 | 190 | 205 |
| 4 | 71 | 89 | 107 | 125 | 142 | 160 | 178 | 196 | 213 | 231 | 249 |

### For Simply Transmitting Power.

| Diameter of Shaft. | Number of Revolutions per Minute. | | | | | | | | | | |
|---|---|---|---|---|---|---|---|---|---|---|---|
| | 100 | 125 | 150 | 175 | 200 | 233 | 267 | 300 | 333 | 367 | 400 |
| Inches. | H.P. | H.P. | H.P. | H.P. | H.P. | H.P. | H.P. | H.P. | H.P. | H.P. | H.P. |
| 1½ | 6.7 | 8.4 | 10.1 | 11.8 | 13.5 | 15.7 | 17.9 | 20.3 | 22.5 | 24.8 | 27.0 |
| 1⅝ | 8.6 | 10.7 | 12.8 | 15.0 | 17.1 | 20 | 22.8 | 25.8 | 28.6 | 31.5 | 34.3 |
| 1¾ | 10.7 | 13.4 | 16.0 | 18.7 | 21.5 | 25 | 28 | 32 | 36 | 39 | 43 |
| 1⅞ | 13.2 | 16.5 | 19.7 | 23 | 26.4 | 31 | 35 | 39 | 44 | 48 | 52 |
| 2 | 16 | 20 | 24 | 28 | 32 | 37 | 42 | 48 | 53 | 58 | 64 |
| 2⅛ | 19 | 24 | 29 | 33 | 38 | 44 | 51 | 57 | 63 | 70 | 76 |
| 2¼ | 22 | 28 | 34 | 39 | 45 | 52 | 60 | 68 | 75 | 83 | 90 |
| 2⅜ | 27 | 33 | 40 | 47 | 53 | 62 | 70 | 79 | 88 | 96 | 105 |
| 2½ | 30 | 39 | 47 | 54 | 62 | 73 | 83 | 93 | 104 | 114 | 125 |
| 2¾ | 41 | 52 | 62 | 73 | 83 | 97 | 111 | 125 | 139 | 153 | 167 |
| 3 | 54 | 67 | 81 | 94 | 108 | 126 | 144 | 162 | 180 | 198 | 216 |
| 3¼ | 68 | 86 | 103 | 120 | 137 | 160 | 182 | 205 | 228 | 250 | 273 |
| 3½ | 85 | 107 | 128 | 150 | 171 | 200 | 228 | 257 | 285 | 313 | 342 |

## Table of Horse Power which may be transmitted by open Single Belts to pulleys running 100 revolutions per minute, the Diameters of the Driving and Driven Pulleys being equal.

| Diam. Pulley. | Width of Belt in Inches. | | | | | | | |
|---|---|---|---|---|---|---|---|---|
| | 2 | 2½ | 3 | 3½ | 4 | 4½ | 5 | 6 |
| In. | HP | HP | HP | HP | HP | HP | HP | HP |
| 6 | .44 | .54 | .65 | .76 | .87 | .98 | 1.09 | 1.31 |
| 6½ | .47 | .59 | .71 | .83 | .95 | 1.07 | 1.19 | 1.42 |
| 7 | .51 | .64 | .76 | .89 | 1.01 | 1.14 | 1.27 | 1.53 |
| 7½ | .55 | .68 | .82 | .95 | 1.09 | 1.23 | 1.36 | 1.64 |
| 8 | .58 | .73 | .87 | 1.02 | 1.16 | 1.31 | 1.45 | 1.75 |
| 8½ | .62 | .77 | .93 | 1.08 | 1.24 | 1.39 | 1.55 | 1.86 |
| 9 | .65 | .82 | .98 | 1.15 | 1.31 | 1.48 | 1.64 | 1.97 |
| 9½ | .69 | .86 | 1.04 | 1.21 | 1.39 | 1.56 | 1.74 | 2.08 |
| 10 | .73 | .91 | 1.09 | 1.27 | 1.45 | 1.63 | 1.81 | 2.18 |
| 11 | .8 | 1. | 1.2 | 1.4 | 1.6 | 1.8 | 2. | 2.4 |
| 12 | .87 | 1.09 | 1.31 | 1.53 | 1.75 | 1.97 | 2.18 | 2.62 |
| 13 | .95 | 1.18 | 1.42 | 1.65 | 1.89 | 2.12 | 2.36 | 2.83 |
| 14 | 1.02 | 1.27 | 1.52 | 1.77 | 2.02 | 2.27 | 2.53 | 3.05 |
| 15 | 1.09 | 1.36 | 1.64 | 1.91 | 2.19 | 2.46 | 2.73 | 3.29 |
| 16 | 1.16 | 1.45 | 1.74 | 2.03 | 2.32 | 2.61 | 2.91 | 3.48 |
| 17 | 1.24 | 1.55 | 1.85 | 2.16 | 2.47 | 2.78 | 3.09 | 3.70 |
| 18 | 1.31 | 1.64 | 1.96 | 2.29 | 2.62 | 2.95 | 3.27 | 3.92 |
| 19 | 1.39 | 1.73 | 2.07 | 2.42 | 2.76 | 3.11 | 3.45 | 4.14 |
| 20 | 1.45 | 1.82 | 2.18 | 2.55 | 2.91 | 3.27 | 3.64 | 4.36 |
| 21 | 1.52 | 1.91 | 2.29 | 2.67 | 3.05 | 3.44 | 3.82 | 4.58 |
| 22 | 1.66 | 2. | 2.4 | 2.8 | 3.2 | 3.6 | 4. | 4.8 |
| 23 | 1.67 | 2.09 | 2.51 | 2.93 | 3.35 | 3.75 | 4.18 | 5.02 |

| Diam. Pulley. | Width of Belt in Inches. | | | | | | | |
|---|---|---|---|---|---|---|---|---|
| | 4 | 5 | 6 | 8 | 10 | 12 | 14 | 16 |
| In. | HP | HP | HP | HP | HP | HP | HP | HP |
| 24 | 3.5 | 4.4 | 5.2 | 7. | 8.7 | 10.5 | 12.2 | 14. |
| 25 | 3.6 | 4.5 | 5.5 | 7.3 | 9.1 | 10.9 | 12.7 | 14.5 |
| 26 | 3.8 | 4.7 | 5.7 | 7.6 | 9.5 | 11.3 | 13.2 | 15.1 |
| 27 | 3.9 | 4.9 | 5.9 | 7.8 | 9.8 | 11.8 | 13.7 | 15.6 |
| 28 | 4.1 | 5.1 | 6.1 | 8.1 | 10.2 | 12.2 | 14.3 | 16.3 |
| 29 | 4.2 | 5.3 | 6.3 | 8.4 | 10.5 | 12.6 | 14.8 | 16.9 |
| 30 | 4.4 | 5.4 | 6.6 | 8.7 | 10.9 | 13.1 | 15.3 | 17.4 |
| 31 | 4.5 | 5.6 | 6.8 | 9. | 11.3 | 13.5 | 15.8 | 18. |
| 32 | 4.7 | 5.8 | 7. | 9.3 | 11.6 | 14. | 16.3 | 18.6 |
| 33 | 4.8 | 6. | 7.2 | 9.6 | 12. | 14.4 | 16.8 | 19.2 |
| 34 | 4.9 | 6.2 | 7.4 | 9.9 | 12.4 | 14.8 | 17.3 | 19.8 |
| 35 | 5.1 | 6.4 | 7.6 | 10.2 | 12.7 | 15.3 | 17.9 | 20.4 |
| 36 | 5.2 | 6.5 | 7.8 | 10.5 | 13.1 | 15.7 | 18.3 | 20.9 |
| 37 | 5.4 | 6.7 | 8.1 | 10.8 | 13.5 | 16.2 | 18.9 | 21.5 |
| 38 | 5.5 | 6.9 | 8.3 | 11. | 13.8 | 16.6 | 19.3 | 22.1 |
| 39 | 5.7 | 7.1 | 8.5 | 11.3 | 14.2 | 17. | 19.9 | 22.7 |
| 40 | 5.8 | 7.3 | 8.7 | 11.6 | 14.6 | 17.5 | 20.4 | 23.3 |
| 42 | 6.1 | 7.6 | 9.2 | 12.2 | 15.3 | 18.2 | 21.4 | 24.7 |
| 44 | 6.4 | 8. | 9.6 | 12.8 | 16. | 19.2 | 22.4 | 25.6 |
| 46 | 6.7 | 8.4 | 10. | 13.4 | 16. | 20.1 | 23.4 | 26.8 |
| 48 | 7. | 8.8 | 10.4 | 14. | 17.4 | 21. | 24.4 | 28. |
| 50 | 7.2 | 9. | 10.9 | 14.6 | 18.2 | 21.8 | 25.4 | 29. |
| 54 | 7.8 | 9.8 | 11.8 | 15.6 | 19.6 | 23.6 | 26.4 | 31.2 |
| 60 | 8.8 | 10.8 | 13.1 | 17.4 | 21.8 | 26.2 | 30.6 | 34.8 |

| Diam. Pulley. | Width of Belt in Inches. | | | | | | | |
|---|---|---|---|---|---|---|---|---|
| | 18 | 20 | 22 | 24 | 26 | 28 | 30 | 32 |
| In. | HP | HP | HP | HP | HP | HP | HP | HP |
| 24 | 16 | 17 | 19 | 21 | 23 | 24 | 26 | 28 |
| 30 | 19 | 22 | 24 | 26 | 28 | 31 | 33 | 35 |
| 36 | 24 | 26 | 29 | 31 | 34 | 37 | 39 | 42 |
| 38 | 25 | 28 | 30 | 33 | 36 | 39 | 41 | 44 |
| 40 | 26 | 29 | 32 | 35 | 38 | 41 | 44 | 47 |
| 42 | 28 | 31 | 34 | 36 | 40 | 43 | 46 | 49 |
| 44 | 29 | 32 | 35 | 38 | 42 | 45 | 48 | 51 |
| 48 | 31 | 35 | 38 | 42 | 45 | 49 | 52 | 56 |
| 50 | 33 | 36 | 40 | 44 | 47 | 51 | 54 | 58 |
| 54 | 35 | 39 | 43 | 47 | 50 | 53 | 58 | 62 |
| 60 | 39 | 44 | 48 | 52 | 57 | 61 | 65 | 70 |
| 66 | 43 | 48 | 53 | 58 | 62 | 67 | 72 | 77 |
| 72 | 47 | 52 | 58 | 63 | 68 | 73 | 78 | 84 |
| 78 | 50 | 57 | 62 | 68 | 74 | 80 | 85 | 91 |
| 84 | 55 | 61 | 67 | 73 | 79 | 86 | 91 | 97 |
| 96 | 63 | 70 | 76 | 84 | 90 | 98 | 104 | 112 |
| 120 | 78 | 88 | 96 | 104 | 114 | 122 | 130 | 140 |
| 144 | 94 | 104 | 116 | 126 | 136 | 146 | 156 | 160 |

The horse power of Double Belts is 1.6 of that given in the table.

### Horse Power of Gear Wheels.

To find the horse power that can be safely transmitted by gear wheels, whose face or breadth of teeth is from 2½ to 3 times the pitch: Multiply the square of pitch by the velocity in feet, per second, of gear at pitch line by .6 equals horse power. Example—Required, horse power that can be safely transmitted by a gear 66.84″ diameter by 1¾″ pitch, and making 60 revolutions per minute. 3.0625 multiplied by 17.5 feet, multiplied by .6 equals 32.16 horse power.

# POWER TRANSMITTING TABLES.

## Standard Hoisting Rope

With 19 Wires to the Strand.

### IRON

| Trade No. | Price per foot, in cents. | Diameter. | Circumference, in inches. | Weight per foot, in pounds, of rope with hemp center. | Breaking strain in tons of 2000 pounds. | Proper working load in tons of 2000 pounds. | Circumference of hemp rope of equal strength. | Minimum size of drum or sheave in feet. |
|---|---|---|---|---|---|---|---|---|
| 1 | 100 | 2¼ | 6¾ | 8.00 | 74 | 15 | 15½ | 8 |
| 2 | 78 | 2 | 6 | 6.30 | 65 | 13 | 14½ | 7 |
| 3 | 69 | 1¾ | 5½ | 5.25 | 54 | 11 | 13 | 6½ |
| 4 | 58 | 1⅝ | 5 | 4.10 | 44 | 9 | 12 | 5 |
| 5 | 53 | 1½ | 4¾ | 3.65 | 39 | 8 | 11½ | 4¾ |
| 5½ | 43 | 1⅜ | 4⅜ | 3.00 | 33 | 6½ | 10¼ | 4½ |
| 6 | 36 | 1¼ | 4 | 2.50 | 27 | 5½ | 9½ | 4 |
| 7 | 29 | 1⅛ | 3½ | 2.00 | 20 | 4 | 8 | 3½ |
| 8 | 26 | 1 | 3⅛ | 1.58 | 16 | 3 | 7 | 3 |
| 9 | 20 | ⅞ | 2¾ | 1.20 | 11.50 | 2½ | 6 | 2¾ |
| 10 | 16 | ¾ | 2¼ | 0.88 | 8.64 | 1¾ | 5 | 2½ |
| 10¼ | 14 | ⅝ | 2 | 0.60 | 5.13 | 1¼ | 4½ | 2 |
| 10½ | 12 | 9/16 | 1⅝ | 0.44 | 4.27 | ¾ | 4 | 1¾ |
| 10¾ | 10 | ½ | 1½ | 0.35 | 3.48 | ½ | 3½ | 1½ |
| 10⅞ | 8 | ⅜ | 1¼ | 0.26 | 2.50 | ¼ | 3 | 1 |

### CAST STEEL

| Trade No. | Price per foot, in cents. | Diameter. | Circumference, in inches. | Weight per foot, in pounds, of rope with hemp center. | Breaking strain in tons of 2000 pounds. | Proper working load in tons of 2000 pounds. | Circumference of hemp rope of equal strength. | Minimum size of drum or sheave in feet. |
|---|---|---|---|---|---|---|---|---|
| 1 | 152 | 2¼ | 6¾ | 8.00 | 155 | 31 | .... | 9 |
| 2 | 120 | 2 | 6 | 6.30 | 125 | 25 | .... | 8 |
| 3 | 100 | 1¾ | 5½ | 5.25 | 105 | 21 | 15¾ | 7½ |
| 4 | 80 | 1⅝ | 5 | 4.10 | 86 | 17 | 14½ | 6 |
| 5 | 71 | 1½ | 4¾ | 3.65 | 77 | 15 | 13½ | 5½ |
| 5½ | 60 | 1⅜ | 4⅜ | 3.00 | 63 | 12 | 12¼ | 5¼ |
| 6 | 50 | 1¼ | 4 | 2.50 | 52 | 10 | 11½ | 5 |
| 7 | 41 | 1⅛ | 3½ | 2.00 | 42 | 8 | 10 | 4½ |
| 8 | 34 | 1 | 3⅛ | 1.58 | 33 | 6 | 9¼ | 4 |
| 9 | 27 | ⅞ | 2¾ | 1.20 | 25 | 5 | 8 | 3¾ |
| 10 | 21 | ¾ | 2¼ | 0.88 | 18 | 3½ | 6½ | 3½ |
| 10¼ | 18 | ⅝ | 2 | 0.60 | 14 | 2½ | 5¼ | 3 |
| 10½ | 17 | 9/16 | 1⅝ | 0.44 | 9 | 1½ | 4¾ | 2¾ |
| 10¾ | 15 | ½ | 1½ | 0.35 | 7½ | 1 | 4½ | 2 |

Bessemer and Siemens-Martin Steel Ropes at same price as Iron Ropes.

NOTE.—When made with *Wire Center*, the price per foot is 10 per cent extra.

## Inclined Plane.

For the benefit of those desiring to use wire rope on slopes, inclined planes, etc., we subjoin a table by which the strain produced by any load can easily be calculated. The table gives the strain produced on a rope by a load of one ton of two thousand pounds, an allowance for rolling friction being made. An additional allowance for the weight of the rope will have to be made.

*Example:* For an inclination of 25 feet in 100 feet, corresponding to an angle of 14 1-12 degrees, a load of 2,000 pounds will produce a strain on the rope of 497 pounds, and for a load of 8,000 pounds, the strain on the rope will be $\frac{497 \times 8{,}000}{2{,}000} = 1{,}988$ pounds.

| Elevation in 100 feet. | Corresponding angle of inclination. | Strain in pounds on rope from a load of 2,000 lbs. | Elevation in 100 feet. | Corresponding angle of inclination. | Strain in pounds on rope from a load of 2,000 lbs. |
|---|---|---|---|---|---|
| 5 | 2⅞° | 112 | 95 | 43½° | 1385 |
| 10 | 5½° | 211 | 100 | 45° | 1419 |
| 15 | 8⅓° | 308 | 105 | 46½° | 1457 |
| 20 | 11⅕° | 404 | 110 | 47¾° | 1487 |
| 25 | 14 1/12° | 497 | 115 | 49° | 1516 |
| 30 | 16¾° | 586 | 120 | 50¼° | 1544 |
| 35 | 19⅓° | 673 | 125 | 51½° | 1570 |
| 40 | 21⅝° | 754 | 130 | 52½° | 1592 |
| 45 | 24¼° | 832 | 135 | 53½° | 1614 |
| 50 | 26⅓° | 905 | 140 | 54½° | 1633 |
| 55 | 28⅝° | 975 | 145 | 55½° | 1653 |
| 60 | 31° | 1040 | 150 | 56¼° | 1671 |
| 65 | 33 1/11° | 1100 | 155 | 57¼° | 1689 |
| 70 | 35° | 1156 | 160 | 58° | 1703 |
| 75 | 37° | 1210 | 165 | 58⅘° | 1717 |
| 80 | 38⅔° | 1260 | 170 | 59½° | 1729 |
| 85 | 40½° | 1304 | 175 | 60¼° | 1742 |
| 90 | 42° | 1347 | | | |

A factor of safety of five to seven times should be taken; that is, the working load on the rope should only be one-fifth to one-seventh of its breaking strength. As a rule, ropes for shafts should have a factor of safety of five, and on inclined planes, where the wear is much greater, the factor of safety should be seven.

## To Find the Diameter of a Gear Wheel, the Number of Teeth and Pitch being given.

| Pitch. | 0 | ⅛ | ¼ | ⅜ | ½ | ⅝ | ¾ | ⅞ | Pitch. |
|---|---|---|---|---|---|---|---|---|---|
| 0 | ....... | .03979 | .07958 | .11937 | .15915 | .19894 | .23873 | .27852 | 0 |
| 1 | .31831 | .35810 | .39789 | .43768 | .47746 | .51725 | .55704 | .59683 | 1 |
| 2 | .63662 | .67641 | .71620 | .75599 | .79577 | .83556 | .87535 | .91514 | 2 |
| 3 | .95493 | .99472 | 1.03451 | 1.07430 | 1.11408 | 1.15387 | 1.19366 | 1.23345 | 3 |
| 4 | 1.27324 | 1.31303 | 1.35282 | 1.39261 | 1.43239 | 1.47218 | 1.51197 | 1.55176 | 4 |
| 5 | 1.59155 | 1.63134 | 1.67113 | 1.71092 | 1.75070 | 1.79049 | 1.83028 | 1 87007 | 5 |
| 6 | 1.90986 | 1.94965 | 1.98944 | 2.02923 | 2.06901 | 2.10880 | 2.14859 | 2.18838 | 6 |
| Pitch. | 0 | .125 | .25 | .375 | .5 | .625 | .75 | .875 | Pitch. |

Multiply the number of teeth by the number in the column opposite the pitch. Example.—Required, the diameter of a gear having one hundred and twenty teeth, 1¾″ pitch? Looking down the first vertical column to 1, and then horizontally to ¾, we find the number 55704 x 120 (the number of teeth) = 66.8448, the diameter

## POWER TRANSMITTING MACHINERY

—INCLUDING—

PATENT FRICTION CLUTCH PULLEYS AND COUPLINGS. SHEAVES FOR ROPE TRANSMISSION, PLAIN OR RUBBER LINED. PILLOW BLOCKS, FLOOR STANDS, HANGERS, PATENT RING OILING JOURNAL BOXES, COUPLINGS, ETC., ETC.

Of most modern design and improved construction. Net prices quoted on application.

# THE PELTON WATER WHEEL CO.

## MANUFACTURERS

**SAN FRANCISCO——NEW YORK**

Iron and Steel Shafts
Iron and Steel Spindles
Balanced Pulleys
Split, or Halved Pulleys
Loose-Pulleys
Flange-Pulleys
Guide, or Mule-Pulleys
Cone-Pulleys
Grooved Pulleys
Clutch Pulleys
Friction Pulleys
Plain Hangers
Bracket-Hangers
Adjustable Hangers
Swivel-Hangers
Swivel-Post-Hangers
Adjustable Floor-Stands
Journal-Boxes
Pillow-Blocks

### Price-List of Shafting, Couplings, Hangers, Boxes and Collars

| | | Diameter of Shaft | $1\frac{7}{16}$ | $1\frac{11}{16}$ | $1\frac{15}{16}$ | $2\frac{3}{16}$ | $2\frac{7}{16}$ | $2\frac{11}{16}$ | $2\frac{15}{16}$ | $3\frac{3}{16}$ | $3\frac{11}{16}$ | $3\frac{15}{16}$ |
|---|---|---|---|---|---|---|---|---|---|---|---|---|
| **Shafting,** | Turned, Straightened and polished. | Price per ft | $0 51 | $0 70 | $0 93 | $1 06 | $1 32 | $1 60 | $2 03 | $2 41 | $2 81 | $3 69 |
| **Couplings,** | **Flange,** finished, fitted and keyed to shafts. | " each pair | 6 17 | 6 33 | 7 65 | 8 93 | 10 63 | 13 18 | 15 73 | 18 70 | 21 25 | 27 63 |
| **Couplings,** | **Clamp,** finished and fitted to shafts | " each | 4 32 | 4 88 | 6 19 | 7 50 | 9 00 | 11 07 | 13 13 | 15 38 | 18 00 | 24 00 |
| **Hangers,** | Double braced adjustable pivoted boxes. | " each 12 in.drop | 3 62 | 4 05 | 5 10 | 6 38 | 7 45 | 9 35 | | | | |
| | | " each 14 in.drop | 3 91 | 4 33 | 5 39 | 6 66 | 7 87 | 9 92 | | | | |
| | | " each 16 in.drop | 4 11 | 4 54 | 5 67 | 6 95 | 8 30 | 10 35 | 12 33 | | | |
| | | " each 18 in.drop | 4 25 | 4 68 | 5 95 | 7 32 | 8 72 | 10 63 | 12 75 | 14 88 | 19 13 | 22 95 |
| | | " each 20 in.drop | | | 6 24 | 7 51 | 9 15 | 11 05 | 13 13 | 15 45 | 19 70 | 23 52 |
| | | " each 24 in.drop | | | 6 80 | 8 30 | 10 20 | 12 55 | 14 67 | 17 03 | 21 90 | 25 93 |
| **Boxes,** | Pillow block, babbitted and faced | " each | 1 90 | 2 15 | 2 90 | 3 50 | 4 25 | 5 15 | 6 25 | 6 75 | 9 15 | 11 65 |
| **Collars,** | Fitted with Set Screws | " " | 0 65 | 0 80 | 1 00 | 1 35 | 1 50 | 1 70 | 1 90 | 2 10 | 2 35 | 2 80 |

The above prices for Shafting are for lengths in stock. Cutting off extra and keyseating extra. Coupling keyseats not charged when Couplings are ordered.

Plain Bearings
Flat Boxes
Wall-Boxes
Wall-Brackets
Flange-faced Couplings
Part Clamp-Couplings
Clutch-Couplings
Friction Couplings
Tight and Loose-Collars
Clamp-Collars
Spur-Wheels
Bevel Wheels
Mitre-Wheels
Fly-Wheels
Chain-Wheels
Rope-Wheels
Drums
Step-Pots
Steps

Steel and Iron Shafts of any length and diameter furnished promptly. For Diameters greater than those noted in above list, special prices will be quoted on application.

# PRICE LIST OF BELTING.

## OAK TANNED LEATHER BELTING.

PRICES PER RUNNING FOOT.

| SIZE | PRICE | SIZE | PRICE | SIZE | PRICE | SIZE | PRICE |
|---|---|---|---|---|---|---|---|
| 1 inch | $0 12 | 5 inches | $0 76 | 17 inches | $2 80 | 34 inches | $ 6 50 |
| 1¼ inch | 16 | 5½ inches | 84 | 18 inches | 3 00 | 36 inches | 7 00 |
| 1½ inch | 20 | 6 inches | 92 | 19 inches | 3 20 | 40 inches | 7 80 |
| 1¾ inch | 24 | 6½ inches | 1 00 | 20 inches | 3 40 | 44 inches | 8 60 |
| 2 inches | 28 | 7 inches | 1 08 | 21 inches | 3 60 | 48 inches | 9 40 |
| 2¼ inches | 32 | 8 inches | 1 24 | 22 inches | 3 80 | 50 inches | 9 80 |
| 2½ inches | 36 | 9 inches | 1 40 | 23 inches | 4 00 | 52 inches | 10 20 |
| 2¾ inches | 40 | 10 inches | 1 56 | 24 inches | 4 20 | 56 inches | 11 00 |
| 3 inches | 44 | 11 inches | 1 72 | 25 inches | 4 40 | 60 inches | 11 80 |
| 3¼ inches | 48 | 12 inches | 1 88 | 26 inches | 4 60 | 64 inches | 12 60 |
| 3½ inches | 52 | 13 inches | 2 04 | 27 inches | 4 80 | 68 inches | 13 40 |
| 3¾ inches | 56 | 14 inches | 2 20 | 28 inches | 5 00 | 72 inches | 14 40 |
| 4 inches | 60 | 15 inches | 2 40 | 30 inches | 5 50 | | |
| 4½ inches | 68 | 16 inches | 2 60 | 32 inches | 6 00 | | |

DOUBLE BELTING TWICE THE PRICE OF SINGLE.

## RUBBER BELTING.

PRICE PER RUNNING FOOT.

| WIDTH. | 2-ply, PER FOOT. | 3-ply, PER FOOT. | 4-ply, PER FOOT. | 5-ply, PER FOOT. | 6-ply, PER FOOT. | 7-ply, PER FOOT. | 8-ply, PER FOOT. |
|---|---|---|---|---|---|---|---|
| 1 inch | $0.07 | ......... | ......... | ......... | ......... | ......... | ......... |
| 1¼ inches | .09 | ......... | ......... | ......... | ......... | ......... | ......... |
| 1½ inches | .11 | $0.13 | ......... | ......... | ......... | ......... | ......... |
| 2 inches | .15 | .17 | $0.21 | ......... | ......... | ......... | ......... |
| 2½ inches | .18 | .22 | .26 | ......... | ......... | ......... | ......... |
| 3 inches | .22 | .26 | .31 | ......... | ......... | ......... | ......... |
| 3½ inches | .26 | .30 | .37 | ......... | ......... | ......... | ......... |
| 4 inches | .30 | .34 | .42 | ......... | ......... | ......... | ......... |
| 4½ inches | .33 | .39 | .47 | ......... | ......... | ......... | ......... |
| 5 inches | .36 | .43 | .52 | ......... | ......... | ......... | ......... |
| 6 inches | .43 | .52 | .62 | ......... | ......... | ......... | ......... |
| 7 inches | .51 | .60 | .73 | ......... | ......... | ......... | ......... |
| 8 inches | .59 | .70 | .84 | $1.05 | $ 1.26 | ......... | ......... |
| 9 inches | .67 | .80 | .95 | 1.18 | 1.42 | ......... | ......... |
| 10 inches | .75 | .90 | 1.07 | 1.33 | 1.60 | ......... | ......... |
| 11 inches | .83 | 1.00 | 1.18 | 1.47 | 1.77 | ......... | ......... |
| 12 inches | .91 | 1.08 | 1.30 | 1.62 | 1.95 | ......... | ......... |
| 13 inches | 1.00 | 1.18 | 1.42 | 1.77 | 2.13 | ......... | ......... |
| 14 inches | 1.08 | 1.28 | 1.54 | 1.92 | 2.31 | ......... | ......... |
| 15 inches | 1.16 | 1.38 | 1.66 | 2.07 | 2.49 | ......... | ......... |
| 16 inches | 1.25 | 1.50 | 1.78 | 2.22 | 2.67 | ......... | ......... |
| 18 inches | 1.41 | 1.70 | 2.02 | 2.52 | 3.03 | $ 3.53 | $ 4.04 |
| 20 inches | 1.58 | 1.90 | 2.26 | 2.82 | 3.39 | 3.95 | 4.52 |
| 22 inches | 1.76 | 2.12 | 2.52 | 3.15 | 3.78 | 4.41 | 5.04 |
| 24 inches | 1.96 | 2.36 | 2.80 | 3.50 | 4.20 | 4.90 | 5.60 |
| 26 inches | ......... | 2.60 | 3.08 | 3.85 | 4.62 | 5.39 | 6.16 |
| 28 inches | ......... | 2.84 | 3.36 | 4.20 | 5.04 | 5.88 | 6.72 |
| 30 inches | ......... | ......... | 3.64 | 4.55 | 5.46 | 6.37 | 7.28 |
| 32 inches | ......... | ......... | 3.92 | 4.90 | 5.88 | 6.86 | 7.84 |
| 34 inches | ......... | ......... | 4.20 | 5.25 | 6.30 | 7.35 | 8.40 |
| 36 inches | ......... | ......... | 4.48 | 5.60 | 6.72 | 7.84 | 8.96 |
| 38 inches | ......... | ......... | 4.76 | 5.95 | 7.14 | 8.33 | 9.52 |
| 40 inches | ......... | ......... | 5.04 | 6.30 | 7.56 | 8.82 | 10.08 |
| 42 inches | ......... | ......... | 5.32 | 6.65 | 7.98 | 9.31 | 10.64 |
| 44 inches | ......... | ......... | 5.60 | 7.00 | 8.40 | 9.80 | 11.20 |
| 46 inches | ......... | ......... | 5.88 | 7.35 | 8.82 | 10.29 | 11.76 |
| 48 inches | ......... | ......... | 6.16 | 7.70 | 9.24 | 10.78 | 12.32 |
| 50 inches | ......... | ......... | 6.44 | 8.05 | 9.66 | 11.27 | 12.88 |
| 52 inches | ......... | ......... | 6.72 | 8.40 | 10.08 | 11.76 | 13.44 |

All Belting guaranteed to be of best quality and highest grade on the market.

Prices on Special Leather and Rubber Belts, as also Discounts on above list, will be quoted on application.

## DOUBLE GATE VALVES, BRASS MOUNTED.

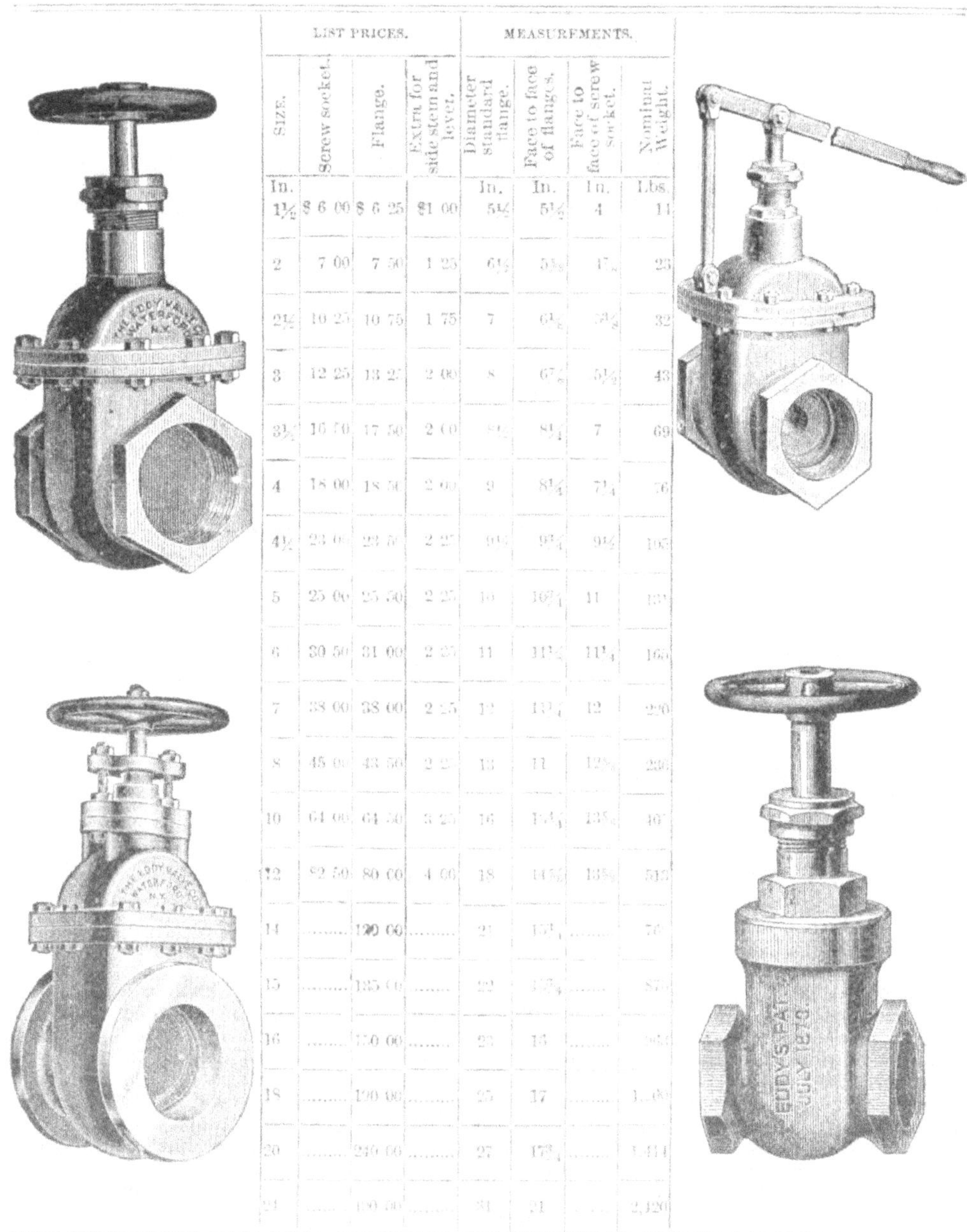


| | LIST PRICES. | | | MEASUREMENTS. | | | |
|---|---|---|---|---|---|---|---|
| SIZE. | Screw socket. | Flange. | Extra for side stem and lever. | Diameter standard flange. | Face to face of flanges. | Face to face of screw socket. | Nominal Weight. |
| In. | | | | In. | In. | In. | Lbs. |
| 1½ | $ 6 00 | $ 6 25 | $1 00 | 5½ | 5½ | 4 | 11 |
| 2 | 7 00 | 7 50 | 1 25 | 6½ | 5⅝ | [illegible] | 23 |
| 2½ | 10 25 | 10 75 | 1 75 | 7 | 6⅛ | [illegible] | 32 |
| 3 | 12 25 | 13 25 | 2 00 | 8 | 6⅞ | 5½ | 43 |
| 3½ | 16 50 | 17 50 | 2 00 | 8½ | 8¼ | 7 | 69 |
| 4 | 18 00 | 18 50 | 2 00 | 9 | 8¼ | 7¼ | 76 |
| 4½ | 23 00 | 23 50 | 2 25 | 9½ | 9¾ | 9½ | [illegible] |
| 5 | 25 00 | 25 50 | 2 25 | 10 | 10¾ | 11 | [illegible] |
| 6 | 30 50 | 31 00 | 2 25 | 11 | 11½ | 11¼ | [illegible] |
| 7 | 38 00 | 38 00 | 2 25 | 12 | 13¼ | 12 | [illegible] |
| 8 | 45 00 | 43 50 | 2 25 | 13 | 11 | 12⅝ | [illegible] |
| 10 | 64 00 | 64 50 | 3 25 | 16 | 13¾ | 13⅝ | [illegible] |
| 12 | 82 50 | 80 00 | 4 00 | 18 | 14½ | 13⅝ | [illegible] |
| 14 | ........ | 120 00 | ........ | 21 | 15¼ | ........ | [illegible] |
| 15 | ........ | 135 00 | ........ | 22 | 15¾ | ........ | [illegible] |
| 16 | ........ | 150 00 | ........ | 23 | 16 | ........ | [illegible] |
| 18 | ........ | 190 00 | ........ | 25 | 17 | ........ | [illegible] |
| 20 | ........ | 240 00 | ........ | 27 | 17¾ | ........ | [illegible] |
| 24 | ........ | [illegible] | ........ | 31 | 21 | ........ | 2,120 |

NOTE.—The above valves are made to carry an equal pressure on both sides of gate, are tested to 300 lbs. cold-water pressure, and are warranted to stand a working pressure of from 150 to 250 pounds. All except the larger sizes made with screw or flange connection as may be desired. Valves furnished of larger capacity than here listed for pipe ranging from 21 up to 60 in. diameter, guaranteed to stand any required pressure. FLUME VALVES, also SLUICE and HEAD GATES, AIR VALVES, etc., furnished to suit any requirements.

## VALVES FOR EXTRA HEAVY PRESSURE.

Valves ranging from 1½ in. to 2½ in. in size, made of gun metal and tested to 1,000 and 1,500 pounds cold-water pressure, are carried in stock, and are good for regular working pressures of from 800 to 1,300 pounds. All heavy pressure valves larger than 2½ in. are made of iron, brass mounted, and are generally carried in stock in sizes ranging from 3 in. to 6 in. Larger sizes can be furnished at short notice. All valves are made to stand any pressure required, and where the conditions are especially severe, will be tested to 2,000 pounds or still higher, if necessary.

## HYDRAULIC PRESSURE GAUGES.

5-in. Dial, all brass, reading to 400 pounds..........................................................price, $15 00

Gauges of either smaller or larger diameter of dial and reading from 100 to 10,000 pounds, furnished if desired. These gauges are specially made to suit the requirements of our business, and are guaranteed to be accurate.

# PELTON

# CABLE AND TELEGRAPHIC CODE.

## STANDARD WHEELS—STANDARD MOTORS.

| | |
|---|---|
| 3-foot wheel, complete | Abbais |
| 4-foot wheel, complete | Abcant |
| 5-foot wheel, complete | Abvert |
| 6-foot wheel, complete | Actran |

| | |
|---|---|
| 6-inch Motor | Acment |
| 12-inch Motor | Acolite |
| 15-inch Motor | Adelton |
| 18-inch Motor | Adlent |
| 24-inch Motor | Admont |
| 30-inch Motor | Adobe |

## MULTIPLE NOZZLE WHEELS.

| | |
|---|---|
| 2-nozzle 3-foot wheel, complete | Baclen |
| 2-nozzle 4-foot wheel, complete | Baden |
| 2-nozzle 5-foot wheel, complete | Badger |
| 2-nozzle 6-foot wheel, complete | Baird |
| 3-nozzle 6-foot wheel, complete | Balde |
| 24-inch quintex wheel, complete | Bants |
| 42-inch quintex wheel, complete | Bardin |
| 48-inch quintex wheel, complete | Baston |

## GENERAL INFORMATION.

| | |
|---|---|
| What sized wheel is advised for —— h. p., under —— head? | Caapon |

EXAMPLE.—50 h. p. is wanted under 150 feet head; the inquiry would read thus: Caapon Ladrillo Maltine.

The reply would be, Abcant, meaning that a 4-foot wheel is advised to fill above.

| | |
|---|---|
| The sized wheel you order will not give the power wanted under conditiors named | Cabazon |
| You require more water or a higher head to get power wanted | Cadamas |
| There is not the power called for in the water under head given | Cadstrom |
| Standard wheel not suited to your requirements; advise special wheel | Cadogan |
| Double-nozzle wheel is not required; a single-nozzle will give ample power | Cahunga |
| Data given not sufficient to make an intelligent estimate. Give particulars as to head, power required, amount of water, length and diameter of pipe line | Calabon |
| What sized wheel do you advise from information given? | Calminos |
| After a careful consideration of all the facts, use your best judgment in filling our order | Callahan |
| We advise a deflecting nozzle with governor | Camartic |
| Give as accurately as possible the h. p. wanted | Camjore |
| Await further instructions by letter | Cameron |
| Would advise using two wheels | Caniadas |
| Would advise using —— wheels | Capistran |
| Letter with full data and plans will be sent | Carmentos |
| We are preparing plans and will forward as soon as completed | Cayugon |
| Advise to fill conditions named | Chaparal |
| How soon and on what terms can you send a competent engineer to make examination of water-power and estimate cost of plant? Wire answer | Chidalgo |
| We will send you a competent engineer to make a survey and report on your water project, to place designated, for —— dollars per day and expenses | Cholone |
| Send at once one of your most experienced engineers to place designated in our last, on terms named, requesting him to report, on arrival, to —— | Chromite |
| Our engineers are all engaged; will endeavor to send one in about —— days. Will that answer? | Champos |

## RELATING TO PRICE.

What will be the price of ——? .......... Dacton
Wire price f. o. b., San Francisco of —— .......... Daices
Wire price f. o. b., New York of —— .......... Danher
Price satisfactory; go ahead with order; shipping instructions will be mailed.......... Darrah
Price satisfactory; but do nothing until you receive plans and orders by mail.......... Dathol
The price is as given in catalogue.......... Daulton
The price named is net cash.......... Daylors

## RELATING TO ORDERS.

Enter our order for wheel to develop —— h. p. under —— feet head.......... Fairfax
EXAMPLE.—Enter our order for wheel to develop 140 h. p. under 180 feet head; which would read: Fairfax Lancade Mancos.
Enter order for wheel under —— feet head with water supply of —— cubic feet per minute, maximum.......... Falerno
Make in sections for mule tranportation, no piece to exceed 250 to 300 pounds.......... Faleston
Execute this order with all possible dispatch.......... Falkner
Wire us when ready for shipment.......... Falstaff
Send wood frame, with frame rods and bolts.......... Famoso
Send complete set frame rods and bolts.......... Fargos
Send 12 extra buckets .......... Farquar
Send duplicate set of buckets.......... Farralon
Send—— extra buckets .......... Farslip
Send an old bucket from wheel as a sample.......... Faucher
Make shaft to gauge sent .......... Feeneys
Make shaft —— inches in diameter exact.......... Feglins
Keyway —— wide by —— deep.......... Fentop
Bore wheel to gauge sent.......... Fermos
Send gauge for diameter of shaft .......... Ferguson
Send guage for bore of wheel.......... Fernando
Shall we furnish driving pulley? If so, give diameter and face.......... Fernside
Send driving pulley for wheel shaft of proper size to run counter with pulley —— diameter x —— face at—— revolutions.......... Filtons
Will belt direct from wheel shaft; driving pulley on wheel shaft must be —— diameter x —— face.......... Finchs
Send driving pulley —— x —— .......... Findley
Give speed of counter and diameter of pulley on same.......... Finnell
Give maximum power wheel is to develop.. .......... Firbunt
Wheel shaft must be —— long.......... Fiskner
Send gauge for wheel shaft.......... Fleener
Send sole plates and foundation bolts at once.......... Flonest
Send sole plates and foundation bolts with wrought-iron housing for masonry foundation.......... Florence
Each nozzle of wheel should have an independent stop-gate.......... Floristo
One nozzle on wheel should have a stop-gate.......... Fordyce
Two nozzles on wheel should have stop-gates.......... Fostrim
Furnish deflecting nozzle with rock shaft, levers and quadrant without governor..... Fowton
Wheel shaft should be extended to take driving pulley on each side of wheel.......... Foxen
Place pulley on right-hand side when facing wheel from nozzle side.......... Fredcam
Place pulley on left-hand side when facing wheel from nozzle side.......... Frilstop
Furnish one additional bearing.......... Frisbies
Provide wheel with ring-oiling journals.......... Frohms
Ring-oiling journals will cost —— extra.......... Fyffes

## RELATING TO PAYMENT.

Enter order; remittance made through.......... Gablain
Draw on us, with bill lading attached, through bank.......... Galions
Draw on us, with bill lading attached, through house of .......... Gantoin
Funds will be on hand at completion of order.......... Garnsey
Must establish credit here for amount of order through bank against documents...... Gaviota
Remit for one-third of order; will draw for balance against documents.......... Gelston
You must put us in funds for one-third of amount of order before we can go on with it; will draw against documents for balance .......... Genevra

## RELATING TO GOVERNORS.

Send governor with wheel.......... Hadley
Send governor with deflecting nozzle.......... Hafford
Send governor with motor ordered.......... Harbins
Send governor for ——.......... Hastors
Governor necessary for your purpose.......... Haywood
Governor is not required for your purpose.......... Heinfen
State character of load and probable variation.......... Hoaglin

## RELATING TO WATER SUPPLY.

| | |
|---|---|
| Cubic feet per minute | Iacua |
| We have a maximum supply of —— cubic feet per minute | Ibex |
| We have a minimum supply of —— cubic feet per minute | Idria |
| Supply variable, but can count on an average of —— cubic feet per minute | Igerna |
| What is your maximum supply of water in cubic feet per minute? | Ignasco |
| State quantity of water in miner's inches | Ilwilde |
| We have an average supply of —— miner's inches | Imusdal |
| We can not obtain more water than the amount stated | Indeek |
| If you can not obtain more water, can the head be increased? | Ionata |

## RELATING TO SHIPMENT.

| | |
|---|---|
| Ship by usual conveyance | Jacobas |
| Await shipping instructions | Jægers |
| Must be ready for shipment on the —— | Jaguas |
| Ship on steamer of the —— | Jalama |
| Ship on next steamer | Jamsan |
| Forward by fast freight | Jamul |
| Shipping weight will approximate —— pounds | Jandor |
| Forward by express | Janes |
| What is the earliest date on which you can ship? | Jaqua |
| On what date was your shipment made? | Jasmin |
| Ship by first sailing vessel | Jayhawk |
| Shall we insure your shipment? | Jelleys |
| It will require about 10 days to fill your order | Jencos |
| It will require about 15 days to fill your order | Jensen |
| It will require about 20 days to fill your order | Jericon |
| It will require about 30 days to fill your order | Jerome |
| Quote best freight rate to —— | Jewetta |
| Car-load rate to —— is —— | Jolont |
| Less car-load rate to —— is —— | Juapon |
| Rate to —— is —— per ton, weight or measure | Juston |

## HORSE-POWER FROM 10 TO 2,000.

| | | | | | |
|---|---|---|---|---|---|
| 10 h. p | Labrea | 150 h. p | Leavits | 290 h. p | Liberty |
| 20 " | Lacana | 160 " | Lecuya | 300 " | Lidells |
| 30 " | Lacosta | 170 " | Lefrancs | 320 " | Liegan |
| 40 " | Laddson | 180 " | Leighton | 340 " | Lillis |
| 50 " | Ladrillo | 190 " | Leland | 360 " | Limas |
| 60 " | Lagona | 200 " | Lempton | 380 " | Lincos |
| 70 " | Lagracia | 210 " | Lenstrom | 400 " | Lindale |
| 80 " | Lahonda | 220 " | Leonas | 420 " | Lindross |
| 90 " | Lairds | 230 " | Lerichs | 440 " | Lisbon |
| 100 " | Lajolla | 240 " | Lethent | 460 " | Lithane |
| 110 " | Lachorbo | 250 " | Levpont | 480 " | Litton |
| 120 " | Laporto | 260 " | Lewiston | 500 " | Livement |
| 130 " | Lamande | 270 " | Lexmont | 1000 " | Livsome |
| 140 " | Lancade | 280 " | Leysant | 2000 " | Lizcont |

## HEAD OF WATER IN FEET.

| | | | | | |
|---|---|---|---|---|---|
| 20 ft. head | Mabels | 230 ft. head | Meacham | 580 ft. head | Midson |
| 30 " | Machos | 240 " | Meadins | 600 " | Milpate |
| 40 " | Maclays | 250 " | Medias | 620 " | Milford |
| 50 " | Macus | 260 " | Meehan | 640 " | Milbræ |
| 60 " | Madler | 270 " | Meekers | 660 " | Milner |
| 70 " | Madera | 280 " | Meineck | 680 " | Milsaps |
| 80 " | Madison | 290 " | Melitta | 700 " | Mindem |
| 90 " | Mafost | 300 " | Melrose | 720 " | Mineota |
| 100 " | Mahews | 320 " | Mentone | 740 " | Minturn |
| 110 " | Malaga | 340 " | Meraza | 760 " | Miramar |
| 120 " | Malakoff | 360 " | Mercury | 780 " | Mockton |
| 130 " | Malkonus | 380 " | Merdout | 800 " | Mojeska |
| 140 " | Malstown | 400 " | Merigan | 820 " | Mojave |
| 150 " | Maltine | 420 " | Merles | 840 " | Moneta |
| 160 " | Mampos | 440 " | Merton | 860 " | Montalvo |
| 170 " | Manstrap | 460 " | Mesant | 880 " | Moraga |
| 180 " | Mancos | 480 " | Mesick | 900 " | Morena |
| 190 " | Manvil | 500 " | Mesilla | 925 " | Morley |
| 200 " | Marcel | 520 " | Mesmer | 950 " | Moulton |
| 210 " | Marcuse | 540 " | Mesquit | 975 " | Mowrys |
| 220 " | Maskot | 560 " | Metsom | 1000 " | Munchton |

## RELATING TO PIPE LINE.

| | |
|---|---|
| What diameter pipe will be required for —— h. p. under —— head, the length being —— feet? | Pachapa |
| What is the length of pipe line? | Pacheco |
| What is the diameter of pipe? | Pacolma |
| Is your pipe line already laid? | Paguay |
| Your pipe is too small to carry the required amount of water | Pahute |
| Advise consulting us in regard to pipe | Pajaro |
| What thickness of iron is advised for —— diameter pipe under —— feet head? | Palmos |
| Would you advise? | Pampas |
| Slip-joint pipe | Panoche |
| Collar and sleeve-joint pipe | Parais |
| Flanged pipe | Pasiana |
| Wrought-iron welded pipe with flanged joints | Pawneck |
| Wrought-iron welded pipe with lock joints | Pendola |
| Send —— feet of pipe of proper weight and diameter to fill our requirements | Penrose |
| Send the pipe of size and weight which in your judgment is best adapted to our wants | Peralta |
| For approximate estimate of cost of pipe, see price list in catalogue | Perrys |
| Must have a profile of pipe line in order to make an intelligent estimate of size, weight, and cost of pipe | Pescados |
| —— gauge pipe is too light to stand the pressure; advise —— gauge | Petrolia |
| Advise using —— feet each, number —— and —— iron | Phoenix |
| Advise pipe of several diameters nested to save freight; shall we make a change? | Pidston |
| Your pipe should be made of —— iron | Piedro |
| No. 18 Birmingham wire gauge | Pinacate |
| No. 16 " " | Pinebaugh |
| No. 14 " " | Placentia |
| No. 12 " " | Pleyton |
| No. 10 " " | Poormas |
| No. 8 " " | Popkins |
| No. 7 " " | Potsdam |
| No. 6 " " | Poways |
| Send pipe cut-punched and formed for mule packing | Preatos |
| Length of section of pipe must not exceed —— feet | Presidio |
| Make pipe in —— sizes to facilitate shipment | Prieton |
| —— pipe will weigh —— pounds per foot | Proberta |
| —— inch pipe, No. —— | Pulcurate |

## RELATING TO LENGTH OF PIPE.

Length of Pipe in Feet.

| | | | | | |
|---|---|---|---|---|---|
| 5 ft. | Sabloa | 200 ft. | Semblos | 2,500 ft. | Sibinto |
| 10 ft. | Sabinal | 250 ft. | Sebranto | 3,000 ft. | Sicorno |
| 20 ft. | Sablato | 300 ft. | Secreto | 3,500 ft. | Sidesto |
| 30 ft. | Sabuyla | 400 ft. | Seculto | 4,000 ft. | Sigrado |
| 40 ft. | Sabinto | 500 ft. | Sedono | 4,500 ft. | Sijando |
| 50 ft. | Sapuisco | 600 ft. | Sedroma | 5,000 ft. | Silhino |
| 60 ft | Sahenta | 700 ft. | Sefranca | 5,500 ft. | Simona |
| 70 ft. | Saspico | 800 ft. | Segona | 6,000 ft. | Sintosa |
| 80 ft. | Satrona | 900 ft. | Seguro | 7,000 ft. | Siperta |
| 90 ft. | Sareilo | 1,000 ft. | Sejaron | 8,000 ft. | Sirosis |
| 100 ft | Sanpete | 1,500 ft. | Sejinto | 9,000 ft. | Sistema |
| 150 ft. | Salesco | 2,000 ft. | Serina | 10,000 ft. | Sitasco |

## RELATING TO DIAMETER OF PIPE.

| | | | | | |
|---|---|---|---|---|---|
| 1 in. diameter | Tacita | 11 in. diameter | Tanka | 26 in. diameter | Tecian |
| 2 in. " | Tacuba | 12 in. " | Tanque | 28 in. " | Teloan |
| 3 in. " | Tacpeto | 13 in. " | Tapias | 30 in. " | Teltos |
| 4 in. " | Talaca | 14 in. " | Tasco | 32 in. " | Tecpan |
| 5 in. " | Talcoch | 15 in. " | Teabo | 34 in. " | Tecuala |
| 6 in. " | Talpa | 16 in. " | Tepon | 36 in. " | Tejano |
| 7 in. " | Tamah | 18 in. " | Teculi | 38 in. " | Tejeria |
| 8 in. " | Tampico | 20 in. " | Tecuma | 40 in. " | Temaxo |
| 9 in. " | Tamos | 22 in. " | Tecute | 44 in. " | Temoson |
| 10 in. " | Tamuin | 24 in. " | Tecax | 48 in. " | Tempoal |

NOTE.—The above supersedes all previous codes, and can be used in conjunction with the A. B. C. and Lieber's Code.

## USEFUL INFORMATION.

1 FOOT = 12 inches = .3058 meters.
1 METER = 3.28 feet = 39.37 inches.
1 CUBIC FOOT = 7.48 gallons = 28.3 liters = .0283 cubic meter.
1 CUBIC METER = 1,000 liters = 264 gallons = 35.32 cubic feet.
1 GALLON = 134 cubic feet = 3.78 liters = .00378 cubic meters.
1 LITER = .001 cubic meter = .035 cubic foot = .264 gallons.
1 CUBIC FOOT OF WATER weighs about 62.4 lbs.
1 CUBIC METER OF WATER weighs about 2,205 lbs.
1 GALLON OF WATER weighs about 8.34 lbs.
1 LITER OF WATER weighs about 2.2 lbs
PRESSURE OF ONE ATMOSPHERE = 14.7 lbs.

PRESSURE IN POUNDS PER SQUARE INCH multiplied by 2.3 will give the vertical head necessary to produce such a pressure.

A VERTICAL HEAD OF WATER in feet multiplied by .434 will give the pressure in pounds per square inch.

A FOOT POUND is one pound raised one foot high in one minute.

A HORSE-POWER is 33,000 pounds raised one foot high in one minute.

TO FIND THE VELOCITY in feet per second necessary to carry a given quantity of water in a pipe of given diameter, divide the quantity in cubic feet per second by the area of the pipe in square feet. The quotient will give the required velocity.

TO FIND THE AREA OF A REQUIRED PIPE, the quantity and velocity being given, divide the quantity in a stated time by the velocity in the same period; the quotient will be the required area, from which the diameter may readily be calculated.

A CONVENIENT METHOD OF CALCULATING the horse-power required to elevate water, is the following:—

$$\frac{\text{Cubic feet per minute} \times \text{feet raised}}{528} = \text{H. P.}$$

IF WATER IS MEASURED BY GALLONS, substitute 3,956 for 528 in the denominator. An allowance of from 25 to 30 per cent should be made for friction.

FRICTION OF WATER IN PIPES increases approximately as the square of the velocity. Doubling the diameter of a pipe increases its capacity four times.

TO FIND THE THICKNESS OF A PIPE to resist safely a given internal pressure, divide the ultimate cohesion of the material in pounds per square inch by the factor of safety. The quotient is the safe cohesive strength of the metal. Divide the given pressure by this safe cohesion. Call the quotient m. To half of m add 1. Multiply the sum by m. Multiply the product by the radius of the pipe in inches.

The above applies to wrought-iron welded pipes. For single-riveted pipe the thickness should be at least 1.8 times that of wrought iron: double-riveted pipe should be about 1.5 thicker than wrought iron.

## RULES FOR CALCULATING SPEED OF PULLEYS.

The diameter of the driver being given, to find its number of revolutions:—

RULE.—Multiply the diameter of the driver in inches by the number of its revolutions, and divide the product by the diameter in inches of the driven; the quotient will be the number of revolutions of the driven.

The diameter and revolutions of the driver being given, to find the diameter of the driven that shall make any given number of revolutions in the same time:—

RULE.—Multiply the diameter in inches of the driver by its number of revolutions and divide the product by the number of revolutions of the driven· the quotient will be the diameter in inches.

To ascertain the size of the driver:—

RULE.—Multiply the diameter in inches of the driven by the number of revolutions you wish it to make, and divide the product by the revolutions of the driver; the quotient will be the diameter of the driver in inches.

www.ingramcontent.com/pod-product-compliance
Lightning Source LLC
Chambersburg PA
CBHW081139170526
45165CB00008B/2727
*9781548110000*